1st International Conference on Machine Learning and Data Engineering (iCMLDE2017)

Sydney, Australia
20-22 November 2017

ISBN: 978-1-7138-0850-3

Printed from e-media with permission by:

Curran Associates, Inc.
57 Morehouse Lane
Red Hook, NY 12571

Printed with permission by Curran Associates, Inc. (2020)

For permission requests, please contact iCMLDE 2017
at the address below.

ICMLDE 2017

Phone: 61 431 070 618
Email: secretary@icmlde.net.au

icmlde.net.au

Additional copies of this publication are available from:

Curran Associates, Inc.
57 Morehouse Lane
Red Hook, NY 12571 USA
Phone: 845-758-0400
Fax: 845-758-2633
Email: curran@proceedings.com
Web: www.proceedings.com

TABLE OF CONTENTS

Proc. 1st International Conference on Machine Learning and Data Engineering (iCMLDE2017)
20-22 Nov 2017, Sydney, Australia

Clustering on Manifolds

Junbin Gao

Discipline of Business Analytics, The University of Sydney Business School
The University of Sydney

Corresponding author's e-mail: junbin.gao@sydney.edu.au

Abstract: *The geometry of a given space characterizes the proximity between data and plays a key role in machine learning. The traditional methods of simply and naively treating data spaces as "flat" Euclidean's may not offer desired effect in variety of learning tasks. In this talk, I would like to report some of my recent research on clustering tasks over manifolds and try to give an introduction to the state-of-the-art learning on manifolds. The focus will be on the low-rank representation (LRR) models on the Grassmann manifold and the curves manifold used in learning tasks from computer vision and time series..*

Keywords: *Machine Learning, Low-Rank Representation and Grassmann manifold.*

Proc. 1st International Conference on Machine Learning and Data Engineering (iCMLDE2017) 20-22 Nov 2017, Sydney, Australia

A Cumulative Timeline Feature Selection Approach for Data Classification

Rafiqul Islam
School of Computing, Charles Sturt University

mislam@csu.edu.au

Abstract

This paper presents a novel feature vector generation technique of malware data which retains high classification accuracy over an extended time period. The proposed approach is to combine the features and accumulating these features with intervals over time. Experimental results show that the proposed method maintains constant classification accuracy and with a standard deviation of 0.92 over the extended time period. These results strongly support the hypothesis that it is possible to develop any classification strategy that will work well into the future.

Key Words: Timeline, Cumulative, Malware, Integrated.

1. INTRODUCTION

Malware authors use various obfuscation techniques to transform a malicious program into undetectable variants with the same core functionalities of the parent malware program. The study in [1] investigates malicious attacks on several websites by creating web honey pots and collecting website-based malware executables over a period of five months. In their study, they collect and analyse malware samples using 6 different antivirus programs, and conduct the same experiment four months later using the updated versions of the 6 programs to determine their efficacy. This work demonstrates that, with training on older malware, some anti-virus software can significantly improve detection rates.

Many researchers have argued that any classification strategy which has been successful in a given time period will not work at a much later date due to changes the data characteristics of nature of the data in particular malicious data. This philosophy is supported by the work in [2 - 11] which indicates that current techniques failed to find a distinctive pattern of malicious data which can be used to identify future malicious data. The argument is that malware evolves with time and eventually becomes unrecognizable from the original form; in addition completely new malware is designed which is unlike any known malware and so would not be detected by anti-virus software constructed to detect known types of malware. In fact, the assumption that malware completely unlike earlier malware is being designed on a major scale is known to be false as indicated by the statistics in [9] showing that barely 25 % of data in 2006 is not a variant of known malware data.

Despite the strong support in the literature for the idea that current detection methods will not easily detect future malware, in this paper, we demonstrate that it is possible to develop a malware detection strategy which retains high accuracy over an extended time period. Therefore, this research provides a significant outcome to classify the future malware data.

2. EXPERIMENTAL SET-UP

2.1 Timeline data preparation

The date of a malware data was that associated with the file when the file was collected. We exported all files, along with their dates, into our database and based on the dates broke the data into groups

To generate groups of malware for use in the testing, we begin with the earliest malware and add month by month across the timeline until all data are grouped. As the first data group, MG_1, we take the earliest-dated 10% of the files. The second data group, MG_2, comprises the data collected immediate after for MG_1, and so on. In all, this results in N malware data groups which are labeled $MG_1 \dots MG_N$. Figure 1 indicates the spread of malware across the N groups with each bar corresponding to a group.

Throughout the test, the set of cleanware files is treated as a single group CG. However, when it is tested against a particular malware group, depending on the comparative size of the two groups, the cleanware group may be divided into subgroups.

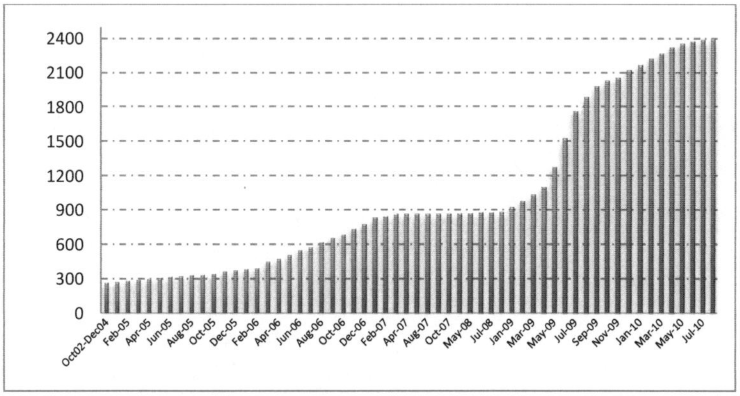

Figure 1. Number of malware executables accumulated by date.

2.2 Test data preparation:

In test data preparation we follow the guidelines in [10-13], use an equal portion of malware and of cleanware data. Figure 2 shows the data preparation process. The selected malware group MG_i is compared with CG. If $|MG_i|$ is smaller than $|CG|$ then we compute the integer part of $|CG|/|MG_i|$ and the integer reminder $0 \leq R < |MG_i|$ as in

$$|CG| = k|MG_i| + R, \text{ for some positive integer k.}$$

We then divide CG into k disjoint groups of equal size. If R > 0, then the remaining elements must be padded out to a (k+1)'st group CG_{k+1}. However, if R = 0, this set is empty and is not used.

If $|MG_i|$ is bigger than $|CG|$ then we compute the integer part of $|MG_i|/|CG|$ and proceed in the same way. This procedure is repeated for every malware group.

2.3 The WEKA interface

In our classification process, we input the feature vectors into the WEKA classification system [6] for which we have written an interface. In all experiments, 10-fold cross validation is applied to ensure a thorough mixing of the features. In this procedure, we first select one group of malware data from a particular data set and divide it into ten portions of equal size; then we select cleanware data of the same size as the group of malware data and also divide it into ten portions. The portions are then tested against each other.

To establish the training set, our detection engine takes nine portions from each of the malware and cleanware to set up the training set and the remaining portions from both malware and cleanware are used for the testing set. As is customary, the training set is used to establish the model and the testing set is used to validate it. The whole process is repeated so that every portion of both malware and cleanware is chosen as testing data; the results are then averaged. In order to ensure that the input vectors are trained and tested over a broad spectrum of classifiers, we chose the following four classifiers from WEKA as they represent differing approaches to statistical analysis of data: Support Vector Machines (SVM), Random Forest (RF), Decision Table (DT) and IB1.

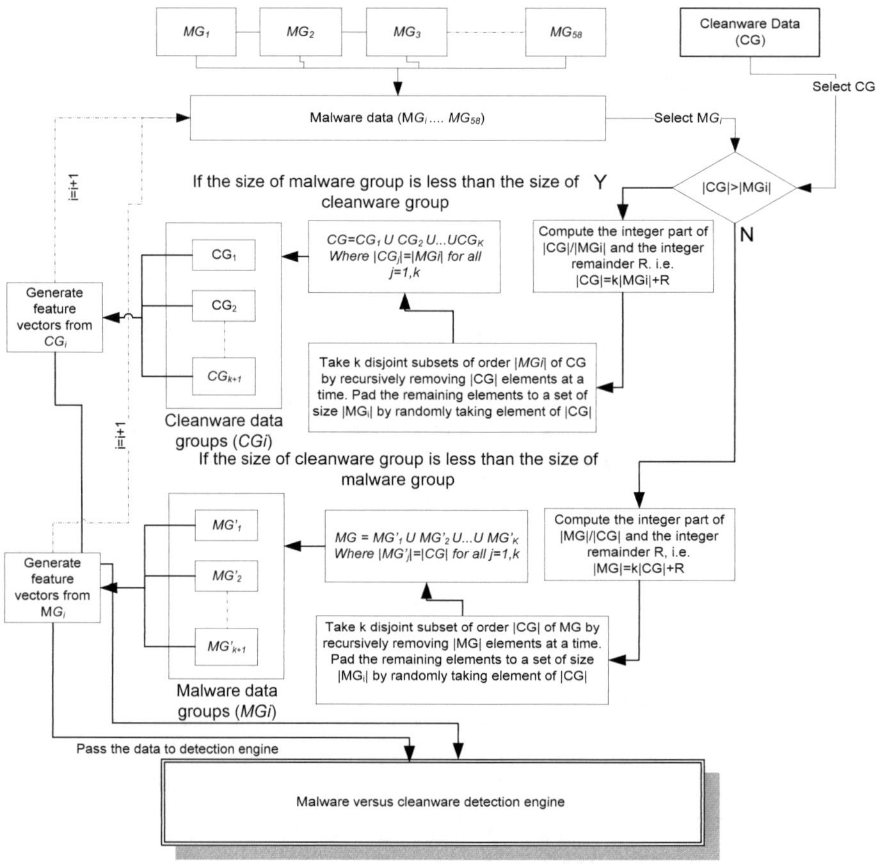

Figure 2. Feature generation

3. EXPERIMENTS AND RESULTS

We ran the entire experiment using each of the four base classifiers SVM, IB1, DT and RF. In addition, each test was run five times and the results averaged in order to ensure that any anomalies in the experimental set-up were discounted.

Figure 3 shows the average results over the timeline data (this is the average of 5 separate tests for each data-group). This average manages to stay above the 80% mark over most of the time period, but drifts under in the immediate previous year of data for some classifiers. As expected, the classifier RF is best on average.

In order to claim that we have a reasonable malware detector over the eight year period, we now focus on the change in accuracy over this time. We therefore compute the standard deviation of the accuracy data from the N values for each of the feature sets and for each of the classifiers, to determine the variation from the mean in each case. Table 1 presents the standard deviation data for the twelve cases. We can see that the Dynamic test shows consistent, and good, performance for all classifiers except IB1; excluding IB1, the difference in spread of the remaining three results is 0.75 points. This is by far the best of our results; however, note that the IB1 result for the dynamic test is worse than all results for the PSI test and worse than two results for FLF. FLF, as expected, has the worst standard deviations, but not all values are worse than those for the PSI test: RF performs slightly better for FLF than SVM does for PSI. The difference in spread of standard deviations for FLF is 1.86 points. There is insufficient support here for removing FLF as a test of the presence of malware. Interestingly, the PSI test has consistent results but with fairly large standard deviations; the difference in spread is 1.49 points.

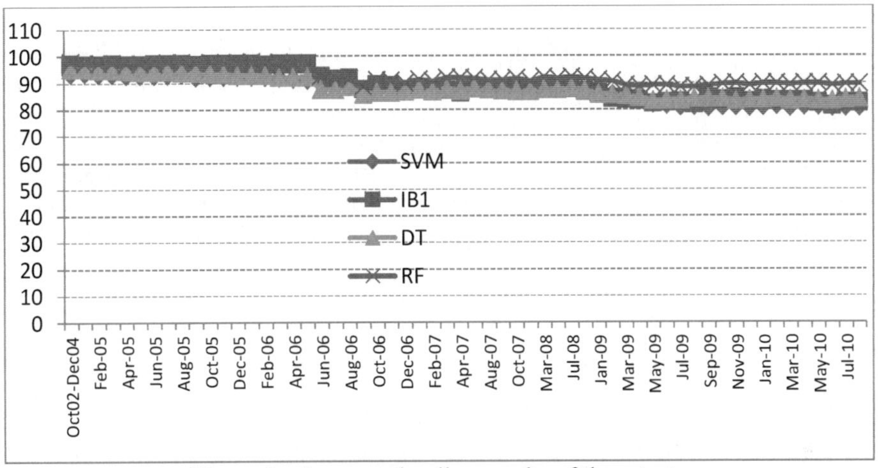

Figure 3. Average timeline results of three tests.

Table 1. Comparison of standard deviations

Feature set	SVM	IB1	DT	RF
FLF	7.422030	5.854992	7.228582	5.755763
PSI	5.849069	5.341312	5.684248	4.354974
Dynamic	0.921473	6.902749	1.668326	1.289221

Turning to an analysis of which classifier is the best, IB1 clearly gives the most consistent results (the difference in spread being 1.56 points), but all three standard deviations are much larger than we would want in a malware detector, and so one of our conclusions would be to exclude this particular classifier from future analysis work on malware. In this case, RF is the best remaining classifier for the FLF experiment, and while RF is not as good as SVM for the dynamic test, it gives better results than the DT classifier for this test, and so we would highly recommend retaining RF in future malware detection analysis work.

4. CONCLUSIONS AND FUTURE WORK

In this paper we have presented a cumulative timeline feature vector generation approach and demonstrated that it retains high accuracy over an extended period of time. The results presented in

Section III indicate that our method retains consistent accuracy over the eight year period. Our approach to feature collection is novel in that we accumulate the features over time segments of an eight year span. In progressively adding additional malware over the time period, we thereby strengthen the accuracy of the test. The implication for anti-virus engines is that they are then able to use previously detected malware to provide features based on which to test new executables.

The results presented in Section III, indicate that no one feature type is the most significant over the eight year span. In our experiment, the integrated features are shown to act independently, and each contributes value to the analysis. However, one conclusion of the discussion in Section III would be to exclude the IB1 classifier from future analysis work on malware, but to retain RF. Therefore, it is expected that combining static and dynamic features in an integrated manner could give a better detection rate; we will explore this in our future work.

REFERENCES

[1.] I. You and K. Yim. Malware obfuscation techniques: A brief survey. In *Broadband, Wireless Computing, Communication and Applications (BWCCA), 2010 International Conference*, pp. 297–300. IEEE, 2010.

[2.] M. Fossi, E. Johnson, T. Mack, D. Turner, J. Blackbird, M. K. Low, T. Adams, D. Mckinney, S. Entwisle, M. P. Laucht, C. Wueest, P. Wood, D. Bleaken, G. Ahmad, D. Kemp, and Samnani. Symantec global internet security threat report trends for 2009, technical report, 2009.

[3.] H. Tang, B. Zhu, and K. Ren. A new approach to malware detection. *Advances in Information Security and Assurance*, pp. 229–238, 2009.

[4.] P. Barford and V. Yegneswaran. An inside look at botnets. *Malware Detection*, pp. 171–191, 2007.

[5.] M. Bailey, J. Oberheide, J. Andersen, Z. Mao, F. Jahanian, and J. Nazario. *Automated Classification and Analysis of Internet Malware*, chapter Recent Advances in Intrusion Detection, pp. 178–197. 2007.

[6.] Mark, H. , Eibe F., Geoffrey H., Bernhard, P., Peter R., Ian H. W. (2009); The WEKA Data Mining Software: An Update; *SIGKDD Explorations*, Volume 11, Issue 1.9

[7.] R. Tian, L.M. Batten, and S.C. Versteeg. Function length as a tool for malware classification. In *Proceedings of the 3rd International Conference on Malicious and Unwanted Software: MALWARE 2008*, pp. 69–76, 2008.

[8.] J. Lee, C. Im, and H. Jeong. A study of malware detection and classification by comparing extracted strings. In *Proceedings of the 5th International Conference on Ubiquitous Information Management and Communication*, p. 75. ACM, 2011.

[9.] M. Braverman, J. Williams, and Z. Mador. Microsoft security intelligence report January-June 2006. Accessed on 19th May 2011 CYVEILLANCE.

[10.] V.P. Nair, H. Jain, Y.K. Golecha, M.S. Gaur, and V. Laxmi. Medusa: Metamorphic malware dynamic analysis using signature from api. In *Proceedings of the 3rd international conference on Security of information and networks*, pp. 263–269. ACM, 2010.

[11.] Islam, R., Tian, R., Batten, L.M., Versteeg, S., (2013). Classification of malware based on integrated static and dynamic features. J. Network and Computer Applications 36, 646–656

[12.] R. Tian, L. Batten, R.I Islam, and S. Versteeg. An automated classification system based on the strings of trojan and virus families. In *Proceedings of the 4rd International Conference on Malicious and Unwanted Software: MALWARE 2009*, pp. 23–30, 2009.

[13.] R. Tian, R. Islam, L. Batten, and S. Versteeg. Differentiating malware from cleanware using behavioural analysis. In *Malicious and Unwanted Software (MALWARE), 2010 5th International Conference*, pp. 23 –30, 2010.

Simplified swarm optimization for feasibility-based rules on constrained engineering design optimization

Wei-Chang Yeh[1] and Chia-Ling Huang[2]

[1]Distinguished Professor, Department of Industrial Engineering and Engineering Management National Tsing Hua University, Hsinchu, Taiwan

yeh@ieee.org

[2]Associate Professor, Department of Logistics and Shipping Management Kainan University, Taoyuan, Taiwan

Corresponding author's E-mail: clhuang@mail.knu.edu.tw

Abstract: *The constrained optimization benchmarks can successfully formulate various practical engineering designs. Therefore, there are various algorithms have been applied to optimize the engineering design by the formulation of the constrained optimization benchmarks since the last decades. The value of the solutions of any objective function that is obtained by any algorithm is ensured only if the solutions of the constraints are falling in feasible solutions. Therefore, it is very important to make sure the solutions are feasible that is the constraints are satisfied before declaring the objective solution obtained by any algorithm. In this work, we use a population-based swarm intelligence algorithm, called simplified swarm optimization (SSO) to optimize the constrained engineering design benchmarks based on the feasible solutions. For the purpose to evaluate the optimization performance, the SSO has performed on three well known constrained engineering design benchmarks including two different types of minimization constrained benchmarks and one engineering constrained and mechanical design benchmark. The computational results have been compared favourably with those obtained using existing algorithms in the literatures. The comparison results demonstrate the proposed SSO well optimize the benchmarks with feasible solutions compared to the other considered algorithms.*

Keywords: simplified swarm optimization (SSO), constrained engineering design, optimization

1. INTRODUCTION

The constrained optimization benchmarks can successfully formulate various practical engineering designs. Therefore, there are various algorithms, such as mine blast algorithm (MBA), harmony search (HS), genetic algorithm (GA), evolutionary programming (EP), improved genetic algorithm (IGA), cultured differential evolution (CA-DE), co-evolutionary particle swarm optimization (CPSO), hybrid particle swarm optimization (HPSO), Nelder-Mead simplex search and particle swarm optimization (NM-PSO), quantum-behaved particle swarm optimization (Q-PSO), and Combining multiobjective optimization with differential evolution (C-DE), have been applied to optimize the engineering design by the formulation of the constrained optimization benchmarks since the last decades [1-15].

As we know, the optimization benchmarks in the real-world practice must be with the constraints. That is all of the optimization engineering design benchmarks are subjected under various constraints. One of the most important techniques is to handle constraints because optimization value obtained by any algorithm seriously losses if the solutions of constraints fall into infeasible regions. Therefore, constraint handling is one of the major concerns when applying the algorithms to solve constrained engineering design optimization benchmarks. First priority, this paper proposes to exam the constraints for feasible solutions or infeasible solutions before comparison the objective solution obtained among each algorithm. Then, the real value of the objective solution obtained by any algorithm can be ensured if having the feasible solutions of constraints.

In this paper, the simplified swarm optimization (SSO) algorithm that belongs to the population-based swarm intelligence is proposed to optimize constrained engineering design benchmarks based on feasible solutions. SSO algorithm is developed and first introduced by Yeh [16] in 2009. SSO has been successfully exploited in considerable studies and applied in various fields. Among these, numerical varieties of constrained engineering design problems have been optimized by SSO, such as optimization of the Disassembly Sequencing Problems [17-19], the series-parallel redundancy allocation problems [20], forecasting wind power [21], and reliability redundancy allocation problems [22].

To demonstrate the performance of SSO, three benchmarks including two different types of minimization constrained benchmarks and one engineering constrained and mechanical design benchmark are considered and demonstrated in this study. The constrained solutions are examined for feasible or infeasible first to ensure the real value of the objective solutions obtained by each algorithm. The experimental solutions are compared with those of previously developed algorithms in literature, the results indicate the proposed SSO performs well for handling feasible solutions of constraints.

2. SIMPLIFIED SWARM OPTIMIZATION (SSO)

SSO is a novel developed soft computing algorithm and first introduced by Yeh [16]. The SSO makes use of a new update mechanism, shown in Eq. (1). The parameter C_w=c_w, C_p= C_w+c_p, and C_g= C_p+c_g, where c_w, c_p, c_g, and c_r are the values of probability; $\rho \in Uniform(0,1)$ and $x \in Uniform(l_j, u_j)$ where l_j and u_j are the lower bound and upper bound in the jth generation for j=1, 2, …, t; solution X_i^t = $(x_{i1}^t, x_{i2}^t, \ldots, x_{i,N_{gen}}^t)$ is a compromise of the current i^{th} solution $X_i^{t-1} = (x_{i1}^{t-1}, x_{i2}^{t-1}, \ldots, x_{iN_{gen}}^{t-1})$, *pBest* $P_i=(p_{i1},p_{i2},\ldots p_{i,N_{gen}}) \in \{X_i^1, X_i^2, \ldots, X_i^t\}$ which is a local best such that $F(P_i) \geq F(X_i^j)$, *gBest* $G=(g_1,g_2,\ldots,g_{N_{gen}}) \in \{P_1,P_2,\ldots,P_{N_{sol}}\}$ which is a global best such that $F(G) \geq F(P_i)$ for i=1,2,…,N_{sol}.

$$x_{ij}^t = \begin{cases} x_{ij}^{t-1} & \text{if } \rho \in [0, C_w) \\ p_{ij} & \text{if } \rho \in [C_w, C_p) \\ g_j & \text{if } \rho \in [C_p, C_g) \\ x & \text{if } \rho \in [C_g, 1) \end{cases} \tag{1}$$

PROCEDURE OF SSO

STEP S0. Generate X_i randomly, let t=1, P_i=X_i, and G=P_j, where $F(P_j) = \underset{i}{Max}\{F(P_i)\}$.

STEP S1. Let i=1.

STEP S2. Update X_i based on Eq. (1) and calculate $F(X_i)$.

STEP S3. If $F(X_i)>F(P_i)$, let P_i=X_i; else, go to STEP S5.

STEP S4. If $F(P_i)>F(G)$, then let G=P_i.

STEP S5. If $i<N_{sol}$, then let i=i+1 and go to STEP S2.

STEP S6. If t=N_{gen}, then G is the final solution and halt; otherwise, let t=t+1 and go to STEP S1.

3. FORMULATION OF THREE BENCHMARKS

The formulations of three benchmarks demonstrated in this study are introduced first in this section.

The First Constrained benchmark

This benchmark is introduced by Brakan and Mccormick [23]. Following, several algorithms are applied to solve this benchmark [1-3, 5].

Two variables are identified and the variable vector is given by

$$\mathbf{x} = (x_1, x_2) \tag{2}$$

The formulation of the first constrained benchmark is displayed as follows:

$$\text{Min} \quad f(\mathbf{x}) = (x_1 - 2)^2 + (x_2 - 1)^2 \tag{3}$$

$$\text{Subject to} \quad g_1(\mathbf{x}) = x_1 - 2x_2 + 1 = 0, \tag{4}$$

$$g_2(\mathbf{x}) = -(x_1^2/4) - x_2^2 + 1 \geq 0, \tag{5}$$

$$-10 \leq x_i \leq 10, \quad i = 1, 2 \tag{6}$$

The Second Constrained benchmark

The various algorithms are applied to solve the second benchmark [1-2, 4, 6].

In this case, seven variables are identified: $x_1, x_2, \ldots, x_7$ and the variable vector is given by

$$\mathbf{x} = (x_1, x_2, x_3, x_4, x_5, x_6, x_7) \tag{7}$$

The formulation of the second constrained benchmark is displayed as follows:

$$\text{Min} \quad f(\mathbf{x}) = (x_1 - 10)^2 + 5(x_2 - 12)^2 + x_3^4 + 3(x_4 - 11)^2 + 10x_5^6 + 7x_6^2 + x_7^4 - 4x_6x_7 - 10x_6 - 8x_7 \tag{8}$$

$$\text{Subject to} \quad g_1(\mathbf{x}) = 127 - 2x_1^2 - 3x_2^4 - x_3 - 3x_4^2 - 5x_5 \geq 0, \tag{9}$$

$$g_2(\mathbf{x}) = 282 - 7x_1 - 3x_2 - 10x_3^2 - x_4 + x_5 \geq 0, \tag{10}$$

$$g_3(\mathbf{x}) = 196 - 23x_1 - x_2^2 - 6x_6^2 + 8x_7 \geq 0, \tag{11}$$

$$g_4(\mathbf{x}) = -4x_1^2 - x_2^2 + 3x_1x_2 - 2x_3^2 - 5x_6 + 11x_7 \geq 0, \tag{12}$$

$$-10 \leq x_i \leq 10, \quad i = 1, 2 \tag{13}$$

The Third Constrained benchmark

Kannan and Kramer [24] propose the third constrained benchmark as shown in Fig. 1, a cylindrical vessel with two hemispherical heads is designed at both ends to minimize the fabrication cost. Several algorithms, such as MBA[1], CPSO[6], HPSO[7], GA1[8], GA2[9], NM-PSO[10], Q-PSO[11] and C-DE[12], are applied to solve this benchmark [1, 6-12].

Four variables are identified: T_s (x_1, thickness of the pressure vessel), T_h (x_2, thickness of the head), R (x_3, inner radius of the vessel), L (x_4, length of the vessel without heads):

$$\mathbf{x} = (x_1, x_2, x_3, x_4) = (T_s, T_h, R, L) \tag{14}$$

The variables T_s and T_h are considered as discrete variables multiples of 0.0625 in, and R and L are treated as continuous variables. The formulation is displayed as follows:

$$\text{Min} \quad f(\mathbf{x}) = 0.6224x_1x_3x_4 + 1.7781x_2x_3^2 + 3.1661x_1^2x_4 + 19.84x_1^2x_3 \tag{15}$$

$$\text{Subject to} \quad g_1(\mathbf{x}) = -x_1 + 0.0193x_3 \le 0, \tag{16}$$

$$g_2(\mathbf{x}) = -x_2 + 0.0954x_3 \le 0, \tag{17}$$

$$g_3(\mathbf{x}) = -\pi x_3^2 x_4 - (4/3)\pi x_3^3 + 1,296,000 \le 0, \tag{18}$$

$$g_4(\mathbf{x}) = x_4 - 240 \le 0, \tag{19}$$

$$1 \le x_i \le 99, \quad i = 1, 2 \tag{20}$$

$$1000 \le x_i \le 2000, \quad i = 3, 4 \tag{21}$$

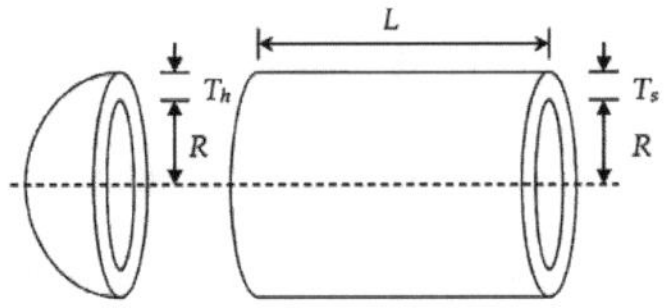

Fig. 1 The pressure vessel design

4. NUMERICAL EXPERIMENTS

The proposed SSO algorithm implemented in all experiments was coded in the C^{++} programming language and run on an Intel Core i7 3.07 GHz PC with 6 GB memory. The experiments used 1000 generations (N_{gen}=1000), the number of solutions was N_{sol}=100.

Experimental results of the first constrained engineering design benchmark are shown in Table 1.

Table 1. Benchmark 1: Comparison of the solutions

	HS [2]	GA[3]	EP[5]	MBA[1]	Bracken and Mccormick [23]	SSO
x_1	0.8343	0.8080	0.8350	0.82288	0.82288	0.822870567
x_2	0.9121	0.8854	0.9125	0.91144	0.91144	0.911435284
$g_1(x)$	5E−03*	3.7E−02*	1.0E−02*	7.05E−09*	7.05E−09*	**0**
$g_2(x)$	5.4E−03	5.2E−02	−7.0E−02	1.73E−08	1.73E−08	6.7309E-06

f(x)	1.3770	1.4339	1.3772	1.3934649	1.393454	**1.39347741**

In the second row of Table 1, the symbol * means the infeasible solution is obtained. The constraint $g_1(x)$=7.05E−09 which is subjected to equal zero as shown in Eq. (4) falls into infeasible solution leads to the value of objective solution *f(x)*=1.393454 obtained by the algorithm proposed by Bracken and Mccormick [23] seriously loses. The constraint $g_1(x)$ also falls into infeasible solution for the other four algorithms HS [2], GA [3], EP [5], and MBA [1]. Therefore, the optimum solution is *f(x)*= **1.39347741** obtained by the proposed SSO algorithm with the best solution obtained at **x**= (0.822870567, 0.911435284), $g_1(x)$=0, and $g_2(x)$= 6.7309E-06 which are all belonging to the feasible solutions.

Experimental results of the second constrained engineering benchmark are shown in Table 2.

Table 2. Benchmark 2: Comparison of the solutions

	IGA[4]	HS[2]	MBA[1]	CA-DE[13]	SSO
x_1	2.330499	2.323456	2.326585	2.330499	2.17932675
x_2	1.951372	1.951242	1.950973	1.951372	1.879634988
x_3	−0.477541	−0.448467	−0.497446	−0.477541	- 0.005798526
x_4	4.365726	4.361919	4.367508	4.365726	4.547563084
x_5	−0.624487	−0.630075	−0.618578	−0.624487	-
x_6	1.038131	1.03866	1.043839	1.038131	0.533158369
x_7	1.594227	1.605348	1.595928	1.594227	1.133152247
$g_1(x)$	4.46E−05	0.208928	1.17E−04	4.46E−05	4.365672767
$g_2(x)$	−252.561723*	−252.878859*	−252.400363*	−252.561723*	0.004486311
$g_3(x)$	−144.878190*	−145.123347*	−144.912069*	−144.878190*	256.0247501
$g_4(x)$	7.63E−06	−0.263414*	1.39E−04	7.63E−06	169.5636351
					32.11470036
f(x)	680.63006	680.6413574	680.6322202	680.6300573	**682.3060712**

In the second row of Table 2, the symbol * means the infeasible solution is obtained. The constraints x_1= −252.561723 and $g_3(x)$= −144.878190 which are subjected to be greater than or equal zero as shown in Eq. (10) and Eq. (11) fall into infeasible solution leads to the value of objective solution *f(x)*= 680.6300573 obtained by the CA-DE algorithm proposed by Becerra and Coello [13] seriously loses. The constraints $g_2(x)$ and $g_3(x)$ also fall into infeasible solution for the other two algorithms IGA [4] and MBA [1]. And the constraints $g_2(x)$, $g_3(x)$ and $g_4(x)$ fall into infeasible solution for the algorithm HS[2]. Therefore, the optimum solution is *f(x)*= **682.3060712** obtained by the proposed SSO algorithm with the best solution obtained at **x**= (2.17932675, 1.879634988, -0.005798526, 4.547563084, -0.533158369, 1.133152247, 4.365672767), $g_1(x)$= 0.004486311, $g_2(x)$= 256.0247501, $g_3(x)$= 169.5636351, and $g_4(x)$= 32.11470036 which are all belonging to the feasible solutions.

Experimental results of the second constrained engineering benchmark are shown in Table 3.

In the second row of Table 3, the symbol * means the infeasible solution is obtained. The constraints x_1= 0.7802 and x_2= 0.3856 which are subjected to be greater than or equal 1 and lesser than or equal 99 as shown in Eq. (20) fall into infeasible solution leads to the value of objective solution $f(x)$= 5889.3216 obtained by the MBA algorithm proposed by Sadollah et al. [8] seriously loses. The constraints x_1 and x_2 also fall into infeasible solution for the other five algorithms GA1[8], GA2[9], CPSO[6], GQ-PSO[11] and C-DE[12]. The constraints x_1, x_2, and $g_3(x)$ fall into infeasible solution for the algorithm HPSO[7]. And the constraints x_1, x_2, and $g_1(x)$ and $g_2(x)$ fall into infeasible solution for the algorithm NM-PSO[10]. Therefore, the optimum solution is $f(x)$= **6008.410926** obtained by the proposed SSO algorithm with the best solution obtained at **x**= (84.99518193, 22.98158617, 172.6715902, 42.63405252), $g_1(x)$= -1.60185E-05, $g_2(x)$= -1.447539968, $g_3(x)$= -0.552476051, and $g_4(x)$= -10.74574313 which are all belonging to the feasible solutions.

Table 3. Benchmark 3: Comparison of the solutions

	GA1[8]	GA2[9]	CPSO[6]	HPSO[7]8	NM-PSO[10]	iQ-PSO[11]	C-DE[12]	MBA[1]	SSO
x_1	0.8125*	0.8125*	0.8125*	0.8125*	0.8036*	0.8125*	0.8125*	0.7802*	84.99518193
x_2	0.4375*	0.4375*	0.4375*	0.4375*	0.3972*	0.4375*	0.4375*	0.3856*	22.98158617
x_3	40.3239	42.0974	42.0913	42.0984	41.6392	42.0984	42.098411	40.4292	172.6715902
x_4	200.0000	176.6540	176.7465	176.6366	182.4120	176.6372	176.637690	198.4964	42.63405252
$g_1(x)$	−3.42E-02	−2.01E−03	−1.37E−06	−8.80E−07	3.65E−05*	−8.79E−07	−6.67E−07	0	-1.60185E-05
$g_2(x)$	−5.28E−02	−3.58E−02	−3.59E−04	−3.58E−02	3.79E−05*	−3.58E−02	−3.58E−02	0	-1.447539968
$g_3(x)$	−304.4020	−24.7593	−118.7687	3.1226*	−1.5914	−0.2179	−3.705123	−86.3645	-0.552476051
$g_4(x)$	−400.0000	−63.3460	−63.2535	−63.3634	−57.5879	−63.3628	−63.362310	−41.5035	-10.74574313
$f(x)$	6288.7445	6059.9463	6061.0777	6059.7143	5930.3137	6059.7208	6059.7340	5889.3216	**6008.410926**

5. CONCLUSION

The SSO algorithm is applied to optimize three constrained engineering design benchmarks. The solutions of the constraints have been examined first as feasible solutions or infeasible solutions before the comparison of the objective solution obtained by each algorithm. Experimental results demonstrate the power of the proposed SSO for dealing with the feasible solutions of constraints. For all of the three constrained benchmarks, the computational results indicate that the proposed SSO obtains feasible solutions for all of the constraints while some infeasible solutions are obtained by the previously developed algorithms in literature [1-13, 23]. The value of the objective solution can be certificated only when the solutions of constraints are feasible solutions. Therefore, the proposed SSO provides optimized objective solutions than other algorithms in terms of examining the solutions of the constraints for feasibility or infeasibility. Moreover, the SSO algorithm can be applied to solve the constrained engineering optimization design especial for which strictly require feasible solutions of the constraints with the accuracy for the objective solution.

Total three famous constrained engineering optimization benchmarks are implemented in this work. In future studies, it will be considered to extend the scale and number of the benchmarks problem.

ACKNOWLEDGMENTS

This research was supported in part by the Ministry of Science and Technology, Taiwan, R.O.C. under grant MOST 104-2221-E-007 -061 -MY3.

REFERENCES

[1] A. Sadollah, A. Bahreininejad, H. Eskandar, M. Hamdi, Mine blast algorithm: A new population based algorithm for solving constrained engineering optimization problems, Applied Soft Computing. 13 (2013) 2592–2612.

[2] K. S. Lee, Z. W. Geem, A new meta-heuristic algorithm for continuous engineering optimization: harmony search theory and practice, Computer Methods in Applied Mechanics and Engineering. 194 (2005) 3902–3933.

[3] A. Homaifar, C.X. Qi, S.H. Lai, Constrained optimization via genetic algorithms, Simulation. 62 (1994) 242–253.

[4] K. Z. Tang, T. K. Sun, J. Y. Yang, An improved genetic algorithm based on a novel selection strategy for nonlinear programming problems, Computers & Chemical Engineering. 35(4) (2011) 615–621.

[5] D.B. Fogel, A comparison of evolutionary programming and genetic algorithms on selected constrained optimization problems, Simulation. 64 (1995) 397–404.

[6] Q. He, L. Wang, An effective co-evolutionary particle swarm optimization for engineering optimization problems, Engineering Applications of Artificial Intelligence. 20 (2006) 89–99.

[7] Q. He, L. Wang, A hybrid particle swarm optimization with a feasibility-based rule for constrained optimization, Applied Mathematics and Computation, 186 (2007) 1407–1422.

[8] C.A.C. Coello, Use of a self-adaptive penalty approach for engineering optimization problems, Computers in Industry. 41 (2000) 113–127.

[9] C.A.C. Coello, E. Mezura Montes, Constraint-handling in genetic algorithms through the use of dominance-based tournament selection, Advanced Engineering Informatics. 16 (2002) 193–203.

[10] E. Zahara, Y.T. Kao, Hybrid Nelder-Mead simplex search and particle swarm optimization for constrained engineering design problems, Expert Systems with Applications. 36 (2009) 3880–3886.

[11] L.D.S. Coelho, Gaussian quantum-behaved particle swarm optimization approaches for constrained engineering design problems, Expert Systems with Applications. 37 (2010) 1676–1683.

[12] Y. Wang, Z. Cai, Combining multiobjective optimization with differential evolution to solve constrained optimization problems, IEEE Transactions on Evolutionary Computation. 16(1) (2012) 117–134.

[13] R.L. Becerra, C.A.C. Coello, Cultured differential evolution for constrained optimization, Computer Methods in Applied Mechanics and Engineering. 195 (2006) 4303–4322.

[14] Jose Ramirez-Marquez, David Coit, Michael Tortorella, A Generalized Multistate-Based Path Vector Approach for Multistate Two-Terminal Reliability, IIE Transactions. 38(6) (2006) 477-488.

[15] Jose Ramirez-Marquez, David Coit, A Monte-Carlo Simulation Approach for Approximating Multistate Two-Terminal Reliability, Reliability Engineering & System Safety. 87(2) (2005) 253-264.

[16] W.C. Yeh, A two-stage discrete particle swarm optimization for the problem of multiple multi-level redundancy allocation in series systems, Expert Systems with Applications. 36(5) (2009) 9192–9200.

[17] W.C. Yeh, H. Lin, A soft computing algorithm for disassembly sequencing, International Conference on Engineering and Computational Mathematics, 2009.

[18] W.C. Yeh, Optimization of the Disassembly Sequencing Problem on the Basis of Self-Adaptive Simplified Swarm Optimization, IEEE TRANSACTIONS ON SYSTEMS, MAN, AND CYBERNETICS—PART A: SYSTEMS AND HUMANS. 42 (1) (2012) 250–261.

[19] W.C. Yeh, Simplified swarm optimization in disassembly sequencing problems with learning effects, Computers & Operations Research. 39 (2012) 2168–2177.

[20] W.C. Yeh, Orthogonal simplified swarm optimization for the series–parallel redundancy allocation problem with a mix of components, Knowledge-Based Systems. 64 (2014) 1–12.
[21] W.C. Yeh, Y.M. Yeh, P.C. Chang, Y.C. Ke, V. Chung, Forecasting wind power in the Mai Liao Wind Farm based on the multi-layer perceptron artificial neural network model with improved simplified swarm optimization, Electrical Power and Energy Systems. 55 (2014) 741–748.
[22] W.C. Yeh, Y. Z. Jiang, V. Y. Y. Chung, X. He, Solving reliability redundancy allocation problems with orthogonal simplified swarm optimization, International Joint Conference on Neural Networks (IJCNN 2015), Killaney, Ireland.
[23] J. Bracken, G.P. Mccormick, Selected Applications of Nonlinear Programming, John Wiley & Sons, New York, 1968.
[24] B.K. Kannan, S.N. Kramer, An augmented lagrange multiplier based method for mixed integer discrete continuous optimization and its applications to mechanical design, Journal of Mechanical Design. 116 (1994) 405–411.

Real-Time Age and Gender Estimation from Face Images

Jang-Hee Yoo, So-Hee Park, and Yongjin Lee

ETRI (Electronics and Telecommunications Research Institute), Daejeon, S. Korea

Corresponding author's e-mail: jhy@etri.re.kr

Abstract: *In this paper, we describe an automated real-time system that estimates age and gender by utilizing a set of facial image sequences from a video camera. The age and gender estimation system consists of four steps: i) detection and extraction of the facial region from input video; ii) selection of the frontal face images from the extracted facial regions using head pose estimation; iii) duplicated face detection and removal by tracking the faces; and iv) age and gender estimation using statistical facial features. Here, LBP features with AdaBoost classifiers are used to detect the face region in a video frame, and the frontal face images are selected using a 3D pose estimation method. In addition, a particle filter-based tracking framework is employed to remove duplicated faces and to improve the accuracy of people counting, and Gabor-LBP features are used to estimate age and gender using a linear SVM and Adaboost classifiers. In experiments, a large number of face datasets are used to train and evaluate the proposed method, and higher performance is achieved in terms of age and gender estimation: 72.53% for age and 98.90% for gender.*

Keywords: Face Attributes, Age and Gender Estimation, Face Recognition.

1. INTRODUCTION

Studies of face recognition have increased due to the large number of application fields, such as user authentication, targeted advertisements, video surveillance and human-robot interaction. Various methods have been developed over the past two decades, and the most widely used methods depend on holistic representation approaches (Bartlett et al., 2002; Etemad et al., 1997; Truk et al., 1991), such as PCA (Principal Component Analysis), LDA (Linear Discriminant Analysis), and ICA (Independent Component Analysis). These approaches are very simple and effective for frontal faces, but are not robust in terms of pose, illumination and expression changes. According to the recent literature, non-statistical learning methods such as Gabor wavelets and LBP (Local Binary Patterns) have been successfully applied in face recognition applications (Gao et al., 2008; Serrano et al., 2010).

On the other hand, age and gender estimation from facial images has also received increasing attention, and many methods have been developed. One major issue in age and gender estimation is how to extract effective representation features from a facial image (Chen et al., 2017). A comprehensive survey of methods and data can be found in (Fu et al., 2010; Han et al., 2015). More recently, deep learning approaches such as CNN (Convolutional Neural Networks) have improved the performance of face recognition and age/gender estimation (Chen et al., 2017). As a commercial application, NEC and Microsoft have produced one of the most practical age and gender estimation systems; semi-supervised learning is employed for age estimation by NEC and deep neural networks by Microsoft.

In this paper, we propose an automated real-time system for age and gender estimation from face images. To achieve this, face detection and pose estimation methods are adopted to acquire frontal face images. Then, LBP and Gabor filters are used to extract face features, and a supervised learning approach using Adaboost is employed to estimate age and gender. To validate the performance of the proposed method, experiments are conducted using a benchmark database. The rest of this paper is organized as follows. Section 2 describes the proposed age and gender estimation system. Experimental results are described in Section 3 which evaluate the proposed method against the benchmark database. Finally, Section 4 concludes this paper, with remarks on future work.

2. REAL-TIME AGE AND GENDER ESTIMATION SYSTEM

Automated age and gender estimation of the human face involves detecting, tracking and normalizing the face in an image sequence. The features extracted from the face image are used to estimate age and gender, and a face tracking method is employed to measure stay time and to count people. Figure 1 illustrates the system architecture used in this study.

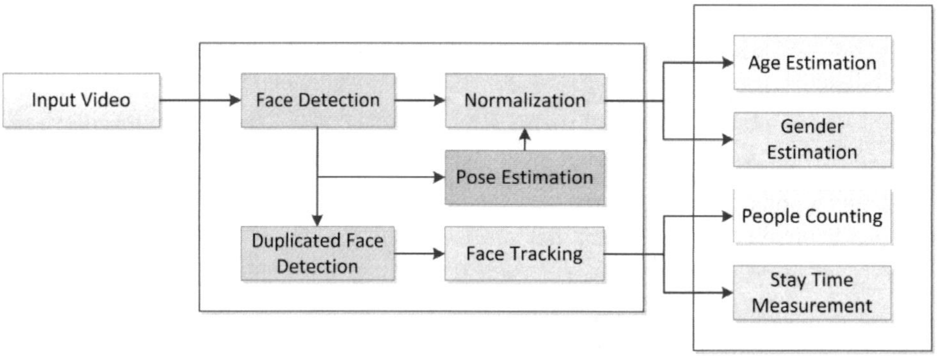

Figure 1. System architecture for age and gender estimation

2.1 Face Detection and Pose Estimation

The LBP-AdaBoost approach (Paris et al., 2011) is used for face detection and eye localization in the input frames. This method adopts the LBP histogram as feature vectors, and the two-class LDA as weak classifiers. LBP features are invariant to monotonic changes in illumination and are computationally efficient. The process of LBP-AdaBoost for face and eye detection is shown in Figure 2. In order to reduce processing time, face detection is performed for only every third input frame, and 64×64 is the minimum size of the face region used for detection. Facial images using localized eye positions are aligned to apply the face estimation algorithm (Park et al., 2011). In addition, an automatic and monocular head pose estimation method is used for acquiring frontal faces using pitch, yaw and roll values. There are many head pose estimation algorithms; however, a point-based algorithm allows fast estimation with no temporal information (Kim et al., 2011). Thus, POSIT (Pose from Orthography and Scaling with ITerations), a hybrid pose estimation algorithm based on both algebraic and optimizing algorithms, is applied for head pose estimation. POSIT has advantages from both approaches and has good speed and accuracy (Kim et al., 2011).

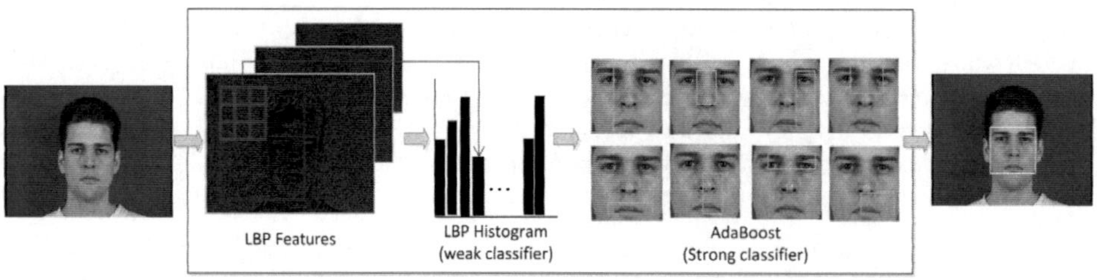

Figure 2. LBP-AdaBoost approach for face detection

2.2 Face Tracking and People Counting

The people counting method based on face tracking consists of face detection, face tracking, tracking failure detection, data association and counting modules. The tracking framework is based on a multi-person tracking algorithm (Choi et al., 2015). An HSV (Hue Saturation Value) histogram is employed as an observation model, and Gaussian perturbation is employed as a motion model in a particle filter-based tracking framework. In order to detect tracking failure, we define a criterion for tracking loss or success as:

$$\mathrm{Status}(k) = \begin{cases} 0 & \text{if } \rho[l_{\hat{\mathbf{s}}_t^k}, l_t^k] < \rho_{\mathrm{th}}, \\ 0 & \text{if } \rho[l_{\hat{\mathbf{s}}_t^k}, l_t^k] - \rho[l_{\hat{\mathbf{s}}_{t-1}^k}, l_{t-1}^k] < \Delta_{\mathrm{th}}, \\ 1 & \text{otherwise}, \end{cases} \tag{1}$$

where $l_{\hat{s}t}^{\;k}$ is the colour histogram constructed at the estimated position of the kth face, l_t^k is the target colour histogram of the kth face, and ρ_{th} and Δ_{th} are predefined threshold values. If a tracking failure is detected by Equation 1, the lost signal (zero) will be transmitted to the face detection module to request a redetection of the lost face. Otherwise, the tracker estimates the state of the current tracking face using particle filtering.

To assign the face detection result to the trackers, a data association method is required. Greedy data association is used to solve the assignment problem, with the matching score defined as:

$$M(\mathbf{s}_{\mathrm{tr}}, \mathbf{s}_{\mathrm{det}}) = \alpha\, p_N(\mathbf{s}_{\mathrm{tr}} - \mathbf{s}_{\mathrm{det}}) + \rho[l_{\mathrm{tr}}, l_{\mathrm{det}}], \tag{2}$$

where $p_N(s_{tr} - s_{det})$ is a Gaussian distribution with a mean of zero and standard deviation σ_d, and α is a weighting factor for the distribution. The face detection result with maximum matching score is associated with the tracker. The associated face detection result and the tracker are then excluded from the matching pool, and this procedure is repeated until no valid pair remains in the matching pool. The proposed method counts the people crossing a pre-configured line with a pre-configured counting direction (e.g. top to bottom, left to right etc.).

2.3 Age and Gender Estimation

After the face and eyes are detected, the facial image can be normalized as a fixed-size image using the localized eye positions. Then, the Gabor-LBP histogram framework (Gao et al., 2008) is used to extract the features for representation power of the spatial histogram (Lee et al., 2012). Using Gaussian derivative filters, the dimensions of the facial features can be reduced by 60% compared with the Gabor wavelet filters. The derived first-order and second-order Gaussian derivative filters are shown in Figure 3. As shown in the figure, the multiple Gaussian derivative filter images are converted by convolving the normalized facial images with the Gaussian derivative filters. Then, each Gaussian derivative filter image is transformed into an LBP map image using the LBP operator, as shown in Figure 3. The LBP patterns at each pixel can be completed by summing the threshold values weighted by a power of two, which characterizes the spatial structure of the local image texture. Finally, each LBP map image is divided into non-overlapping sub-regions with a predefined bin size, and a total of 36 bins are used for each LBP map image. The LBP histograms of the sub-regions are concatenated to form the final histogram sequence as the face representation.

In general, age estimation can be characterized as a multi-class or regression problem, and gender estimation is defined as a two-class problem. Thus, although we can apply AdaBoost directly to gender estimation, applying Adaboost to age estimation is not straightforward. Figure 4 shows a hierarchical approach for age estimation. For example, suppose that we have eight age classes. Then, the training data can be divided into two sub-groups recursively, and Adaboost can be trained such that the two sub-groups are well-discriminated. As shown in the figure, a total of seven classifiers are required to train age data for estimation (Lee et al., 2016).

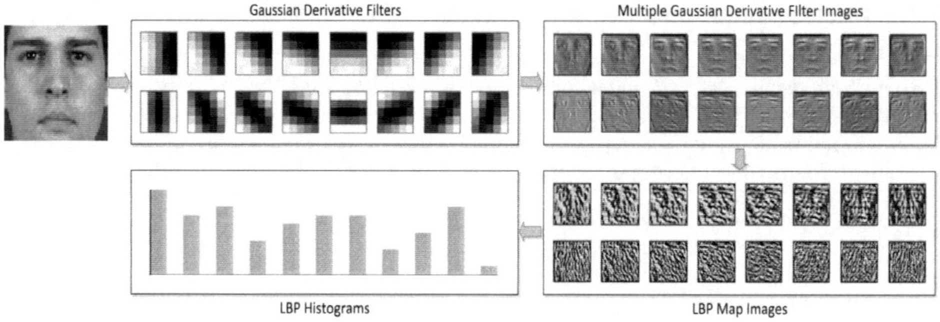

Figure 3. Face features using LBP with Gaussian derivative filters

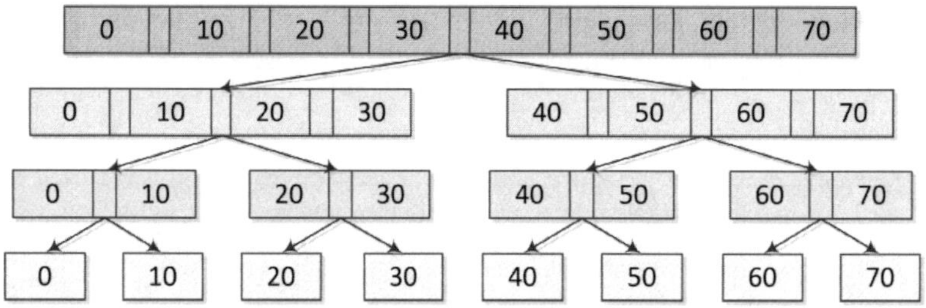

Figure 4. Hierarchical multi-classifier

3. EXPERIMENTAL RESULTS

In the experiments, 25,866 face images (21,392 for training and 4,474 for testing) are used for gender estimation. To estimate gender, Adaboost is trained until the classification for training data reaches nearly 100%. Table 1 shows the test results for the estimation accuracy of testing data. As can be seen from the table, the female dataset shows very good accuracy compared to the male dataset. The reason for this is that many images in the testing data were collected via internet surfing, and internet images are frequently smoothed using image processing tools such as Photoshop. Human features are very sensitive to texture and skin tone, and most of the selected features are located around the meaningful areas for recognition, such as eyebrows, nose, cheekbones and jaw-line. Nevertheless, we obtain a good performance of 98.90% CCR (Correct Classification Rate) in a real-time test.

Table 1. Gender estimation using facial images

Gender	Training (# of Subjects)	Testing (# of Subjects)	Test Result (CCR, %)
Male	12,525	2,092	78.63
Female	8,867	2,382	98.05
Total	21,392	4,474	87.37

We have much less training data for age estimation than for gender classification, and the gathering process of ground truth for age is rather difficult; even humans cannot estimate real age accurately. Thus, the estimation accuracy is low.

Table 2. Age estimation using facial images

Age	0-2	2-7	8-13	14-19	20-29	30-39	40-49	50-59	60-69	70+	Total
Training (# of Sub.)	33	59	114	430	500	500	500	500	239	90	2,965
Testing (# of Sub.)	34	38	33	154	118	72	104	49	30	9	641
Test Results (CCR, %)	85.29	68.42	57.58	61.69	56.78	37.50	44.34	53.06	63.33	77.78	56.32

Table 2 shows the number of experimental datasets and the accuracy for age estimation of the testing data. As can be seen from the table, similar classes are often confused with each other. For example, small children (2~7) are confused with older children (8~13), older children with teenagers (14~19), and thirties (30~39) with forties (40~49). Although the estimation performance is low, the results are quite meaningful. In fact, humans are also easily confused between these age groups. In addition, we obtained 72.53% CCR with a tolerance of ± 5 years, in a real-time test.

4. CONCLUSIONS

We describe a real-time age and gender estimation system with an AdaBoost classifier for facial images. The age and gender estimation system consists of face detection, pose estimation, face tracking and facial feature extraction modules. To estimate age and gender, the extracted facial features are used to train the AdaBoost classifier; people counting and stay time measurement are completed using duplicate face detection with the face tracking method. In a real-time test, estimation rates of 98.90% CCR for gender and 72.53% CCR for age are achieved. These results show promising performance and higher estimation rates than those of an earlier age and gender estimation approach.

ACKNOWLEDGMENTS

This work was supported by Institute for Information & Communications Technology Promotion (IITP) grant funded by the Korea government (MSIT) (No.2017-0-00046, Basic Technology for Extracting High-level Information from Multiple Sources Data base on Intelligent Analysis).

REFERENCES

Bartlett MS, Movellan JR, and Sejnowski TJ (2002). Face Recognition by Independent Component Analysis, IEEE Trans. on Neural Networks, 13(6), pp.1450-1464.

Chen S, Zhang C, Dong M, Le J, and Rao M (2017). Using Ranking-CNN for Age Estimation, in Proc. of the Int. Conf. on Computer Vision and Pattern Recognition (CVPR), pp.742-750.

Choi JW, Moon DS, and Yoo JH (2015). Robust Multi-Person Tracking for Real-Time Intelligent Video Surveillance, ETRI Journal, 37(3), pp.551-561.

Etemad K and Chellappa R (1997). Discriminant Analysis for Recognition of Human Face Images, Journal of the Optical Society of America A: Optics and Image Science, and Vision, 14(8), pp.1724-1733.

Fu Y, Guo G, and Huang TS (2010). Age Synthesis and Estimation via Faces: A Survey, IEEE Trans. Pattern Anal. Intell. 32(11), pp.1955-1976.

Gao T and He M (2008). A Novel Face Description by Local Multi-channel Gabor Histogram Sequence Binary Pattern, in Proc. of the Int. Conf. on Audio, Language and Image Processing, pp.1240-1244.

Han H, Otto C, Liu X, and Jain AK (2015). Demographic Estimation from Face Images: Human vs. Machine Performance, IEEE Trans. Pattern Anal. Intell. 37(6), pp.1148-1161.

Kim WW, Park SH, Hwang JK, and Lee SY (2011). Automatic Head Pose Estimation from a Single Camera using Projective Geometry, in Proc. of the Int. Conf. on Info. Comm. and Signal Processing.

Lee HS, Chung YS, Yoo JH, and Won CH (2012). Face Recognition based on Sparse Representation Classifier with Gabor-Edge Components Histogram, in Proc. SITIS, pp.105-109.

Lee Y, Yoo JH, and Kim KH (2016). Feature Analysis for Human Gender and Age Recognition, Draft.

Paris S, Glotin H, and Zhao Z (2011). Real-time Face Detection using Integral Histogram of Multi-scale Local Binary Patterns, LNCS 6838, Springer-Verlag, pp.276-281.

Park SH, Yoo JH (2012). Robust Eye Localization using Correlation Filter, in Proc. ICCCS, pp.137-140.

Turk MA and Pentland AP (1991). Face Recognition using Eigenfaces, in Proc. CVPR, pp.586-591.

Serrano A, Diego IM, Conde C, and Cabello E (2010). Recent Advances in Face Biometrics with Gabor Wavelets: A Review, Pattern Recognition Letters, 31, pp.372-381.

Proc. 1st International Conference on Machine Learning and Data Engineering (iCMLDE2017) 20-22 Nov 2017, Sydney, Australia

Time Series Anomaly Detection in Online Social Network: Challenges & Solutions

Md Rafiqul Islam[1], Rafiqul Islam[2] and Abu Raihan M Kamal[3]

[1, 3] Islamic University of Technology (IUT), Dhaka, Bangladesh
[2] Charles Sturt University, NSW, Australia
Corresponding author's E-mail: rafiqulislam.cse24@gmail.com

Abstract

In the field of data mining, the social network is one of the complex systems that poses significant challenges in this area. Time series anomaly detection is one of the critical applications. Recent developments in the quantitative analysis of social networks, based largely on graph theory, have been successfully used in various types of time series data. In this survey paper, we review the studies on graph theory to investigate and analyze time series social networks data including different efficient and scalable experimental modalities. We provide some applications, challenging issues and existing methods for time series anomaly detection.

Keywords: Social networks, Time Series Analysis, Anomaly Detection.

1. INTRODUCTION

Social networks have been an upcoming research field over the last few years, and they have gained an established position globally. Social networks provide online hang out space for everybody and using this technology anybody can communicate with their interested friend and share their information, photos, and videos. But this prominent technology also opens the door for unlawful activities. These illicit activities are also alluded to as anomalies. Anomalies emerge in online social networks as a result of specific people, or gatherings of people, rolling out sudden improvements in their examples of connection or communicating in a way that extraordinarily contrasts from their companions. The effects of this strange conduct can be seen in the subsequent network structure. In the field of mathematics, graph theory is a major area to model relations between objects and to represent a connected network structure. From the past decade, researchers are using graph theory to quantify aspects such as similarity, hierarchy and network efficiency of complex network structure in many other fields.

Anomalies are typically defined in terms of deviation from some expected behavior. In general, the definitions of anomalies as "patterns in data that do not conform to a well-defined notion of normal behavior". A record of phenomenon irregularly varying with time is referred to as time series. The analysis of time series is critical for many different scientific areas as often it contains some anomalies. However, it is necessary to carefully examine graphs of the data as a first step of the time series analysis. For example, to experiments the time series analysis of social networks, S. Asur and B. A. Huberman [1] Collected the quantity of tweets about a specific 'point from twitter. Also in their trials, each question is characterized by the Twitter handle of a prevalent artist recorded on the site reverbnation.com, and they recover all tweets that straightforwardly address this craftsman.

Over the past decade, many of the researchers are working to improve their understanding of how the social network work and how they can detect the anomalies. A large number of prominent techniques have been applied to detect time series anomalies in social network. Time series anomaly detection contains many challenging issues. Predetermined number of tasks has been done on applying time arrangement oddity recognition systems to online social networks because of the issues, for example, precision, computational multifaceted nature, and protection, absence of marked datasets and absence of adequate data. Different methods (distance-based, distribution-based, and clustering based.) have been applied to find anomalies in time series data. But these methods do not work well.

The major contributions of this survey paper are as follows:

We aim to familiarize time series anomaly detection in social network. We describe how the anomalies can be identified and how time series anomalies can be detected from social network.

We discuss many of the relevant works that have shown how various anomaly detection techniques of social network have already been used to detect anomalies.

Finally, with these ideas established we propose one technique that can help ameliorate the future directions of social network research.

The remainder of the paper is organized as follows: Section 2 presents the related work of time series anomaly detection in social network. While the analysis of time series anomaly detection in social networks in Section 3. The discussion is presented in Section 4. Finally, the conclusion is provided in Section 5.

2. RELATED WORK

A Social network is an online platform which provides an accessible space for everyone, especially young adults. They use this technology to socialize with absorbed accompany and acquaintances, and to allotment photos, information, and videos. This controlling phenomenon, which captures the structure and dynamics of person-to-technology and person-to-person interaction, is being used for numerous purposes such as education, business, medical, telemarketing, ball and adulterous activities. Social networks have been upcoming research areas over the last few years and they have increased a recognized position globally. In general, Social networks are self-organizing, emergent, and complex, such that a world-wide articular pattern appears from the local alternation of the elements that make up the system. These patterns become added credible as network size increases. Some of the most popular social network websites are Facebook, Google+, LinkedIn, Twitter, Messenger, Viber, WhatsApp, YouTube, WeChat etc. Sometimes this popular social network also opens the door for anomalies. Anomalies in online social networks can announce irregular and generally illegal behavior. Detection of such anomalies has been acclimated to analyze awful individuals, including spammers, animal predators, and online fraudsters [2, 3]. It is truly difficult to take care of an anomaly detection problem in a general frame. In this manner, an anomaly detection method should be developed and modified for a particular application by embracing thoughts from various teaches, for example, insights, machine learning, and data mining.

Enos, et al. [4] has proposed social network analysis to comprehend the interoperability related with the DoD arrangement of frameworks. It applied a few centrality measurements to a system of Major Defence Acquisition Programs (MDAPs) to evaluate the interoperability of individual frameworks inside the arrangement of frameworks. In particular, it inspects the contrasts between the degree, closeness, and an eigenvector centrality measurement to distinguish which metric best speaks to the interoperability of individual frameworks. Gardounis et al.[5] has purposed a SNA applied system to dissect venture forms and their changeability as an outcome of BIM implementation. The proposed point of view features applied endeavours for upgrading and connecting holes apparent in the current methodologies. Such holes prompt a failure to oblige the mind boggling frameworks surfacing among partaking groups and their refined needs, likewise influencing the incorporation of various mechanical interfaces and the coordination of the viable trade of huge scale data.

P. Kazienko and N. Chawla. [6] has presented the idea of the social network classifier that can autonomously order an on-screen character from the relationship among performers. They have tried the likelihood of utilizing the social network classifier as a supporting classifier. It can be utilized to enhance the certainty level of different classifiers paying little heed to their temperament. Himaja, N., and G. Murali [7] has suggested a general straggler-aware execution technique, SAE, to help the assessment transporter inside the cloud. It introduces a novel computational disintegration strategy that causes straggling capacity extraction strategies into more magnificent grained sub systems that are then distributed over groups of PCs for parallel execution. Test comes about demonstrate that SAE can pace up the assessment by method for as much as 1.77 occasions in correlation with best in class arrangements.

Awasthi, Abhishek [8] has designed a Genetic Clustering Algorithm which can be utilized for a) recognizing every one of the bunches in a chart and b) the group of only one given hub. The Genetic algorithm utilized for the grouping issue additionally consolidates the Island demonstrate with movement making the calculation much proficient than a solitary populace Genetic algorithm. The execution of the island show with movement can be upgraded substantially more by parallel programming. Doostmohammadian et al. [9] has built up the vital conditions for circulated discernibleness of social networks modelled as LSI frameworks. They have a portray essential perceptions, specialist order, and system network that empower every operator to deduce any social marvels advancing over a given social digraph. Specifically, they demonstrated that the circulated recognisability requires no a bigger number of perceptions than the brought together case; nonetheless, it requires certain characterization and availability necessities on the specialists watching those states. Cutillo et al. [10] has brook down the connection between the social network graph topology and the achievable security. They have watched three measurements, in particular degree dispersion, bunching coefficient and blending time, and demonstrate that they give key bits of knowledge on the protection level of the OSN.

Patil, Neha An. Et al [11] has actualized and enhanced run time post or message channel for a social network application. Because of enhanced manage set it is demonstrated that different sorts of assaults as for private information of client and malignant social exercises are avoided. They have made a constant application on to break down framework execution and it is discovered its execution is expanded essentially when contrasted with other online networking like Facebook, Twitter, LinkedIn and so forth.; Mathematicians Andrew Beveridge and Jie Shan published Network of Thrones in Math Horizon Magazine where they analyzed graph theory and then applied social network analysis algorithms to the network to find the most important characters in the network and a community detection algorithm to find clusters of characters [12].

K. Saha et al.,[13] has pointed of this investigation were to (1) build a record that deliberate the consciousness of various statistic bunches around schizophrenia-related data on Facebook; (2) consider how this list contrasted crosswise over statistic gatherings and how it corresponded with correlative Web-based (Google Trends) and non– Web-based factors about populace prosperity (emotional well-being markers and framework), and (3) look at the relationship of Facebook determined schizophrenia list with different sorts of online movement and in addition disconnected health and mental health and indicators. K. Garimella et al.[14] has exactly inspected the part of a recently presented Twitter highlight, 'cite retweets' (or 'quote RTs') in political talk, particularly whether it has prompted enhanced, common, and adjusted trade. Quote RTs enable clients to cite the tweet they retweet, while including a short remark. They investigation utilizing substance, system and group marked information demonstrates that the element has expanded political talk and its dispersion, contrasted with existing highlights. They talk about the ramifications of our discoveries in comprehension and lessening on the web polarization.

3. TIME SERIES ANOMALY DETECTION IN ONLINE SOCIAL NETWORK

Time-series data is a type of temporal data which is naturally high dimensional and large in data size. Time-series data are of interest due to their ubiquity in various areas ranging from science, engineering, business, finance, economics, healthcare, to government. While each time-series is consisting of a large number of data points it can also be seen as a single object. Genshiro Kitagawa, [15] described the definition of time series data in his book "Introduction to Time Series Modeling" as follows: A record of phenomenon irregularly varying with time is called time series. He also provided the examples of time series data. Typical time series examples are economic data like stock prices; meteorological data like temperature or rainfall and also medical data.

Anomaly detection is an important problem that has been studied within various research areas including online social networks, fraud, industrial damage, critical systems, image processing insurance, health care, military and bank. Anomaly detection, also named as outlier detection, refers to detecting patterns which do not accede with accepted behaviors. Besides especially anomaly detection is coming to play an increasing great role in social network. Detecting time series anomalies is an important issue in an area as it could be either a sign of a cogent botheration or carrying advantageous advice for the analyzer.

3.1. Application of time series anomaly detection

The analysis of time series anomaly detection on scientific applications can be examined of the diverse fields. Kitagawa [15] points out that it is also essential to carefully examine graphs of the data as a first step of the time series analysis. So it is easier to identify the next step of the analysis and find appropriate strategies for statistical modelling. Some of the important applications of time series anomaly detections are as follows.

	Applications	References
Time series Anomaly Detection	Social network	[16,17]
	Flight sequences data from aircrafts	[18,19]
	Shape of medical data	[20,21]
	Complex brain network	[22,23]
	Outlier light curves of periodic stars	[24,25]
	Eco-system	[26,27]
	Attack detection in recommender systems	28
	Heart beat pulses	29

3.2. Challenges of time series anomaly detection

There are some major challenges related with anomaly detection for time series are:

- There are many different means in which an anomaly occurring in a time series may be identified. An event within a time series may be anomalous; a subsequence within a time series may be anomalous; or an entire time series may be anomalous with respect to a set of normal time series.
- The training and test time series can be of different lengths.
- For detecting anomalous subsequence, the real length of the subsequence is generally unknown.
- Best similarity/distance measures which can be used for different types of time series is not easy to determine. Simple measures like Euclidean distance do not consistently perform well as they are awful acute to outliers and they also cannot be used if the time alternation are of different lengths.
- Analyzing performances of many anomaly detection algorithms are awful affected to noise in the time series data, as detect appropriate anomalies from noise is an arduous task.
- Time series in absolute applications are usually continued and as the length increases the computational complication also increases.

3.3. Solutions of time series anomaly detection

A large number of approaches have been developed to study time series anomaly detection of social networks and each of these technologies aims to aid in examine and assessing the extent of social network. Although these techniques are able to detect anomalies from social network, sometimes they are not able to provide the clear results. They have not the capacity to cover the entire social network rather they provide only information. Moreover, many of the anomaly detection algorithms apprehend multiple time series to be at a comparable scale in consequence while for a lot of the data it is not true.

Method	Description	References
Principal Component Analysis (PCA)	Used for identifing unusual time series in a large collections of time series.	[30]
RLPSVDD (Relaxed of linear programming Support Vector Data Description)	RLPSVDD solves a linear programming problem to provide a flexible data description for time series anomaly detection.	[31]
Decision tree classifier	Automatically extracting and summarizing reports	[32]
Latent Dirichlet Allocation (LDA)	Uses an abnormality estimation scheme based on probabilistic topic modeling and seasonal-trend decomposition to find and examine relevant message subsets.	[33]
Two-stage approach	Used for anomaly detection in large dynamic networks, in a context where in principle any type of anomaly should be detected.	[34]
EFS (Evolving fuzzy ststem)	Automatically analyze the user profiles in real times of a specific community of users. This analysis includes the detection of outliers, the clustering of profiles and their classification.	[35]
Statistical modeling	Used for the early and accurate anomaly detection from the time series of the negative tweets.	[36]
Regression Model	Describe the behavior	[37]
Multiple Linear Regression model	Describe the behavior of random variables	[37]
Autoregressive model	Ability to predict	[38]
Multilevel Regression Model	Examine independent and interactive effects of variables	[39-42]
Gaussian Mixed Model	Support categorical and continues values	[43]
ARIMA	Able to cluster seasonality patterns	[44]
Markov Chain	Multiple variable support	[45]
Hidden Markov Chain	Able to capture the dependencies between variables	[46]

4. DISCUSSION

Nowadays, Time series anomaly detection in the social network is a novel research field globally. Many researchers already engaged with this research field still there are some challenging issues need to be identified. In this paper we have surveyed a number of methods for detecting time series anomalies in social networks. We found that these different methods can be usefully categorized based on characterization of anomalies as being static or dynamic and labelled or unlabeled. Depending on this characterization, different features of the network may be examined, and for this we are proposing Model based clustering algorithm is one of the best technique. A future aspect could be to apply this unique algorithm or framework to further and may be larger datasets to get better insights. Based on these new datasets and algorithm it would also be a good idea to extend the social network properties. As a result of this it would also be possible to compare the determined results better and have more possibilities to find anomalies.

5. CONCLUSION

Time series anomaly detection in social networks have been played an important rule globally. In this survey paper we describe some review of studies and identify some challenging issues of time series analysis. Besides, detecting these challenging problems we propose one novel technique. We hope this technique will be played an increasingly important rule in the evolvement of social network in near future.

REFERENCES

1. S. Asur and B. A. Huberman, "Predicting the future with social media," in Web Intelligence and Intelligent
2. Agent Technology (WI-IAT), 2010 IEEE/WIC/ACM International Conference on, 2010, pp. 492-499.
3. B. A. Huberman, D. M. Romero, and F. Wu, "Social networks that matter: Twitter under the microscope," 2008.
4. J. M. J. M. Gottman, Time-series analysisa comprehensive introduction for social scientists, 1981.
5. J. R. Enos, M. Mansouri, and R. Nilchiani, "Applying social network analysis to systems of systems: The case of the 2016 Puerto Rico power outage," in System of Systems Engineering Conference (SoSE), 2017 12th, 2017, pp. 1-6.
6. F. Gardounis, H.-Y. Chong, and X. Wang, "A conceptual framework for Social Network Analysis of Building Information Modelling in construction projects," in Research and Innovation in Information Systems (ICRIIS), 2017 International Conference on, 2017, pp. 1-6.
7. P. Kazienko and N. Chawla, Applications of Social Media and Social Network Analysis: Springer, 2015.
8. N. Himaja and G. Murali, "Social network analysis for efficient cloud data access," in Communication and Electronics Systems (ICCES), International Conference on, 2016, pp. 1-4.
9. A. Awasthi, "Clustering algorithms for anti-money laundering using graph theory and social network analysis," 2012.
10. M. Doostmohammadian and U. A. Khan, "Graph-theoretic distributed inference in social networks," IEEE Journal of Selected Topics in Signal Processing, vol. 8, pp. 613-623, 2014.
11. L. A. Cutillo, R. Molva, and M. Onen, "Analysis of privacy in online social networks from the graph theory perspective," in Global Telecommunications Conference (GLOBECOM 2011), 2011 IEEE, 2011, pp. 1-5.
12. N. A. Patil and A. S. Manekar, "A Novel Approach to Prevent Personal Data on a Social Network Using Graph Theory," in Computing Communication Control and Automation (ICCUBEA), 2015 International Conference on, 2015, pp. 186-189.
13. A. Beveridge and J. Shan, "Network of thrones," Math Horizons, vol. 23, pp. 18-22, 2016.

14. K. Saha, I. Weber, M. L. Birnbaum, and M. De Choudhury, "Characterizing Awareness of Schizophrenia Among Facebook Users by Leveraging Facebook Advertisement Estimates," Journal of medical Internet research, vol. 19, 2017.
15. K. Garimella, I. Weber, and M. De Choudhury, "Quote RTs on Twitter: usage of the new feature for political discourse," in Proceedings of the 8th ACM Conference on Web Science, 2016, pp. 200-204.
16. G. Kitagawa, "Introducing to Time Series Modeling, Chapman & Hall," ed: USA, CRC Press, 2010.
17. M. C. Chuah and F. Fu, "ECG anomaly detection via time series analysis," in International Symposium on Parallel and Distributed Processing and Applications, 2007, pp. 123-135.
18. S. Akhavan and G. Calva, "'Automatic Anomaly Detection in ECG Signal by Fuzzy Decision Making," in Proceedings of 6th International Conference on Fuzzy Theory and Technology, 1998, pp. 96-98.
19. N. A. Heard, D. J. Weston, K. Platanioti, and D. J. Hand, "Bayesian anomaly detection methods for social networks," The Annals of Applied Statistics, vol. 4, pp. 645-662, 2010.
20. D. Savage, X. Zhang, X. Yu, P. Chou, and Q. Wang, "Anomaly detection in online social networks," Social Networks, vol. 39, pp. 62-70, 2014.
21. V. Rajagopalan and A. Ray, "Symbolic time series analysis via wavelet-based partitioning," Signal Processing, vol. 86, pp. 3309-3320, 2006.
22. D. K. Tolani, M. Yasar, A. Ray, and V. Yang, "Anomaly Detection in Aircraft Gas Turbine Engines," JACIC, vol. 3, pp. 44-51, 2006.
23. M. Bicego and V. Murino, "Investigating hidden Markov models' capabilities in 2D shape classification," IEEE Transactions on Pattern Analysis and Machine Intelligence, vol. 26, pp. 281-286, 2004.
24. Z. Liu, J. X. Yu, L. Chen, and D. Wu, "Detection of shape anomalies: A probabilistic approach using hidden markov models," in Data Engineering, 2008. ICDE 2008. IEEE 24th International Conference on, 2008, pp. 1325-1327.
25. L. Wei, E. Keogh, and X. Xi, "Saxually explicit images: Finding unusual shapes," in Data Mining, 2006. ICDM'06. Sixth International Conference on, 2006, pp. 711-720.
26. P. Malhotra, L. Vig, G. Shroff, and P. Agarwal, "Long short term memory networks for anomaly detection in time series," in Proceedings, 2015, p. 89.
27. E. Bullmore and O. Sporns, "Complex brain networks: graph theoretical analysis of structural and functional systems," Nature reviews. Neuroscience, vol. 10, p. 186, 2009.
28. P. Protopapas, J. Giammarco, L. Faccioli, M. Struble, R. Dave, and C. Alcock, "Finding outlier light curves in catalogues of periodic variable stars," Monthly Notices of the Royal Astronomical Society, vol. 369, pp. 677-696, 2006.
29. U. Rebbapragada, P. Protopapas, C. E. Brodley, and C. Alcock, "Finding anomalous periodic time series," Machine learning, vol. 74, pp. 281-313, 2009.
30. H. Cheng, P.-N. Tan, C. Potter, and S. Klooster, "Detection and characterization of anomalies in multivariate time series," in Proceedings of the 2009 SIAM International Conference on Data Mining, 2009, pp. 413-424.
31. S. Zhang, A. Chakrabarti, J. Ford, and F. Makedon, "Attack detection in time series for recommender systems," in Proceedings of the 12th ACM SIGKDD international conference on Knowledge discovery and data mining, 2006, pp. 809-814.
32. R. J. Hyndman, E. Wang, and N. Laptev, "Large-scale unusual time series detection," in Data Mining Workshop (ICDMW), 2015 IEEE International Conference on, 2015, pp. 1616-1619.
33. C. Huang, G. Min, Y. Wu, Y. Ying, K. Pei, and Z. Xiang, "Time Series Anomaly Detection for Trustworthy Services in Cloud Computing Systems," IEEE Transactions on Big Data, 2017.
34. J. Krumm and E. Horvitz, "Eyewitness: Identifying local events via space-time signals in twitter feeds," in Proceedings of the 23rd SIGSPATIAL International Conference on Advances in Geographic Information Systems, 2015, p. 20.
35. J. Chae, D. Thom, H. Bosch, Y. Jang, R. Maciejewski, D. S. Ebert, et al., "Spatiotemporal social media analytics for abnormal event detection and examination using seasonal-trend

decomposition," in Visual Analytics Science and Technology (VAST), 2012 IEEE Conference on, 2012, pp. 143-152.

36. J. A. Iglesias, A. García-Cuerva, A. Ledezma, and A. Sanchis, "Social network analysis: Evolving Twitter mining," in Systems, Man, and Cybernetics (SMC), 2016 IEEE International Conference on, 2016, pp. 001809-001814.
37. K. Ikeda, G. Hattori, C. Ono, H. Asoh, and T. Higashino, "Early detection method of service quality reduction based on linguistic and time series analysis of twitter," in Advanced Information Networking and Applications Workshops (WAINA), 2013 27th International Conference on, 2013, pp. 825-830.
38. T. Nguyen, D. Phung, B. Dao, S. Venkatesh, and M. Berk, "Affective and content analysis of online depression communities," IEEE Transactions on Affective Computing, vol. 5, pp. 217-226, 2014.
39. M. De Choudhury, S. Counts, and E. Horvitz, "Social media as a measurement tool of depression in populations," in Proceedings of the 5th Annual ACM Web Science Conference, 2013, pp. 47-56.
40. O. L. Haimson, N. Andalibi, M. De Choudhury, and G. R. Hayes, "Relationship breakup disclosures and media ideologies on Facebook," New Media & Society, p. 1461444817711402, 2017.
41. K. Saha, I. Weber, M. L. Birnbaum, and M. De Choudhury, "Characterizing Awareness of Schizophrenia Among Facebook Users by Leveraging Facebook Advertisement Estimates," Journal of medical Internet research, vol. 19, 2017.
42. K. Garimella, I. Weber, and M. De Choudhury, "Quote RTs on Twitter: usage of the new feature for political discourse," in Proceedings of the 8th ACM Conference on Web Science, 2016, pp. 200-204.
43. C. Biernacki, G. Celeux, and G. Govaert, "Assessing a mixture model for clustering with the integrated completed likelihood," IEEE transactions on pattern analysis and machine intelligence, vol. 22, pp. 719-725, 2000.
44. M. Corduas and D. Piccolo, "Time series clustering and classification by the autoregressive metric," Computational Statistics & Data Analysis, vol. 52, pp. 1860-1872, 2008.
45. M. Ramoni, P. Sebastiani, and P. Cohen, "Multivariate clustering by dynamics," in AAAI/IAAI, 2000, pp. 633-638.
46. M. Bicego, V. Murino, and M. A. Figueiredo, "Similarity-based clustering of sequences using hidden Markov models," in International Workshop on Machine Learning and Data Mining in Pattern Recognition, 2003, pp. 86-95.

Proc. 1st International Conference on Machine Learning and Data Engineering (iCMLDE2017)
20-22 Nov 2017, Sydney, Australia

Facial Expression Classification Based on Shape Feature of Emoticons

Akira Fujisawa[1], Kazuyuki Matsumoto[1], Minoru Yoshida[1] and Kenji Kita[1]
[1]Tokushima University, Tokushima, Japan

Corresponding author's E-mail: c501547001@tokushima-u.ac.jp

Abstract

Emoticons are used in the situation of textual communication such as web mails and internet forums. Many of the existing studies dealing with classification or extraction of emoticons regard emoticons as a kind of character string and focus on what characters constitute the emoticons or how they are lined up. However, emoticons are used to express human facial expressions, and characters constituting them represent various facial parts such as eyes, nose, mouth, etc. Such characters can be identified as different facial parts depending on their positions, and facial expressions are thought to be represented by the combinations of their shape features. In this study, we classified the facial expressions of emoticons by focusing on the shape features of those emoticons. To deal with shape features of emoticons, we converted emoticons, which are text data, to image data. Emoticons are mainly formed by line segments of characters, and use only black and white colors. Therefore, other factors such as colors and shades were not considered as the feature to classify the facial expressions. In the experiments, we used image features that did not require color information. As the result of comparative experiment with the 1-nearest neighbor method using character features, the facial expression recognition rate is 52% when using the Histograms of Oriented Gradients(HOG) was used as image feature. By this result, proposed method improved recognition rate by 2 % than using baseline .

Keywords: Facial expression classification, Emoticon, Image feature, k-means clustering.

1. INTRODUCTION

Communication with various type of non-verbal expressions has been increasing on Social Networking Services. Among them, emojis and emoticons are type of expressions that can be easily embedded into sentences. Japanese emoticons are unique in their variety. An emoticon is thought to be a kind of deformation that represents a human face by using character sequence.

The existing engineering studies [Suzuki et al.(2006), Urabe et al.(2013)] have proposed several methods that classify facial expressions from emoticons, that extract emoticons from sentences and that recommend emoticons for users to use. These studies focused on the difference between a word sequence consisting of a sentence and emoticon (a symbol sequence) to extract or classify emoticons. However, an emoticon is different from a simple expression using symbols, it is a picture drawn with characters, that is to say, a kind of Ascii Art. Because emoticon convey visually appealing information, a simple matching between the symbol sequences might fail.

In this study, we extracted image features by treating an emoticon as an image and identifying its shape elements. By using this feature, we thought that we could grasp the characteristics of its shape more accurately as more than a simple sequence of symbols, and thus recognize facial expressions from emoticons more precisely.

Many studies on emoticon have tried to extract information from emoticons, recognize facial expressions from emoticons, recommend emoticons to use, etc. As our study aimed to classify facial expressions of emoticons, we would like to introduce some studies related to facial expression recognition from emoticons.

Yamada et al.(2007) attempted recognition of facial expression in emoticons by training the sequence of characters in those emoticons using N-gram. Ptaszynski et al.(2010) developed a "CAO system" for extracting emoticons from text. Okumura(2016) constructed a large-scale emoticon dictionary. Because many emoticon patterns are possible by combining symbols to be added before and after the emoticon, he proposed a method to extract the primitive form of emoticons.

These exiting studies proposed method to recognize facial expressions of emoticons by constructing emoticon dictionaries to match with the target emoticon, recognizing each emoticon as a sequence of characters/symbols or extracting character frequency from each emoticon as a feature. Their approaches are different from our approach as they cannot consider the similarity of the characters that are similar in shape but otherwise different from each other.

2. FEATURE EXTRACTION FROM EMOTICON

In this paper, we used three features: Local Binary Pattern(LBP), Histograms of Oriented Gradients(HOG) and GIST Descriptor to extract shape features of emoticons. These features are often used for image retrieval and object recognition. In the case of emotion, because there is no need to consider the difference of size between the emoticons and the gradients, our approach does not use Scale-Invariant Feature Transform(SIFT) or Random Sample Consensus(RANSAC) that are robust to scale or angle variation.

2.1. Conversion of Emoticon into Image

This subsection describes a process to convert an emoticon into an image. Emoticons are generally expressed in text form. That means if the kind of font used to display the emoticon is different, there will be some differences. This study uses the "sazanami-mincho" font that is often used in the Japanese language environment. To convert an emoticon into an image, we used an image processing application: ImageMagick[1]. Then, we created an image file (jpeg) by setting the background color as white, the font color as black and the sizes of all emoticons as the same. We extracted three kinds of features from this image file: LBP, GIST and HOG.

2.2. Extraction of Image Features

Fig. 1 shows a flow of extracting image features from an emoticon then registering them into a database with annotation of facial expression labels.

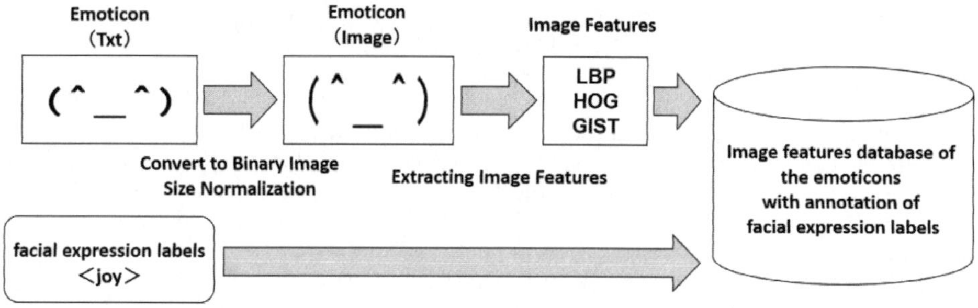

Fig. 1 The flow of construct the image features database of emoticons.

[1] http://imagemagick.org/script/index.php

The image features used in this paper are described below.

2.2.1 Local Binary Pattern(LBP)

Local Binary Pattern (LBP) was proposed by Ojala et al.(1996). The method can extract local features of an image and pattern-based features used for texture analysis. Because the feature is less subject to illumination change and takes low calculation cost, it is sometime used in the field of similar image retrieval.

2.2.2 GIST

Gist, which was proposed by Torralba et al.(2001), is one of the global image features obtained from the whole image. It has been reported that the method using Gist feature can achieve high performance in a task of detecting a copied image. Gist divides the whole image into 4x4 areas and describes scene information by applying the Gabor filter of various directions or frequencies for each area. It is also often used in a task of recognizing a scene.

2.2.3 Histograms of Oriented Gradients (HOG)

The Histograms of Oriented Gradients (HOG) feature, which was proposed by Dalal et al.(2005), is used for object recognition as a gradient-based feature. The method creates a histogram by calculating the gradient direction of brightness over a certain area. Although the HOG feature is not invariant to rotation or scale change, is can detect a person with high accuracy.

2.3. Baseline Feature

We used the character frequency vector and character string itself (character-based feature) as the baseline features. The character frequency vector is a feature that divides an emoticon into characters and then counts the frequency of each character. The similarity among character frequency vectors is calculated by cosine similarity. When the character-based feature is used, an emoticon is treated as a character string and the Levenshtein distance (LD) is calculated.

3. FACIAL EXPRESSION CLASSIFICATION METHOD

Many existing studies related to emoticons classify facial expressions based on six basic expressions proposed by Ekman et al.(1971). Our study aims to classify eight facial expressions including these six basic facial expressions (joy, hate, anger, sorrow, fear (anxiety) and surprise) in addition to 'shame' and 'none' We extracted features from emoticons and created a facial expression classification model by a machine learning method.

The proposed method uses k-Nearest Neighbor Method to check which emoticon is most similar in shape to another emoticon by observing the similarity degree between the extracted features. Then, we propose a method to extract image features from each character that forms an emoticon, not from the total characters of the emoticon itself. The similarities of each part or part order can be considered by this method.

The method calculates the Levenshtein distance between the sequences that are sorted from the image features extracted from each character in an emoticon. Each feature is extracted from the inputted emoticon, then the similarities (cosine similarity, Levenshtein distance) and the features of the emoticons in the training data are compared. Finally, the outputted emoticons are sorted according to the order of similarity and the top *k* emoticons are obtained. The facial expressions ranked on the top *k* emoticons are counted and outputted as the facial expressions of the inputted emoticon.

4. EXPERIMENT

To compare the performance of the proposed method and that of the baseline method, we conducted the evaluation experiment. The following subsection describes the experimental condition. Then, we introduce the experimental results.

4.1. Experimental Condition

4.1.1 Dataset

The dataset used in the experiment consists of the emoticons expressing facial expression that are selected from Kaomoji café(2016), Kaomoji.ru(2012), Kaomoji Copipe(2016), and FACEMARK PARTY(1999).
The dataset are normalized after being removed as parts of the facial contour. The detail of the dataset is shown in Table 1.

Table 1 Experimental dataset

Facial expression	joy	fear	surprise	angry	sorrow	none	hate	shame
Number	431	395	297	189	158	155	138	127

4.1.2 Evaluation Index

We compared the performances of the proposed method and the baseline method by using several evaluation indexes. If the class outputted by the classifier was the same as the correct class, the result was regarded as correct. Then recall rate, precision rate, and F1-score were calculated for each facial expression. We compared, analyzed and discussed the result by using these numerical values as the evaluation indexes. **Error! Reference source not found.** shows the combinations of features and similarities. We used the dataset for the 10-hold cross-validation experiment.

Table 2 Combinations of features and similarities

Method	Feature	Target	Similarity
Baseline	Character frequency	Overall of emoticon	Cosine
	Character string		LD
Proposed method	LBP		Cosine
	HOG		
	GIST		
	LBP	Character basis	LD
	HOG		
	GIST		

4.1.3 Evaluation Experiment

Experimental programs were constructed by Python2.7. To extract HOG, LBP, we used library "scikit-image[2]". To extract GIST, we used lear-gist-python[3].

[2] scikit-image(http://scikit-image.org/docs/dev/api/skimage.html)
[3] lear-gist-python(https://github.com/tuttieee/lear-gist-python/tree/master/sample)

4.2 Experimental Result

In the case that k equal 1, the F1-score of the proposed method and the baseline method are shown in Table 3 and Table 4. The proposed method using the HOG feature obtained high F1-scores, on average, without large differences depending on the facial expressions. Although the LBP feature was calculated with low complexity, it could not classify enough because it was too simple. Because GIST is a global feature, it might have been difficult to distinguish emoticons.
On the other hand, F1-scores by the baseline method differed greatly depending on the facial expressions in both the using character vector and the character string as feature and decreased as k increased. The baseline method might have been significantly affected by the distribution of facial expressions of the emoticons.

Table 3 F1-scores of experimental result which target is overall of emoticons(k=1)

	joy	fear	surprise	angry	sorrow	none	hate	shame
LBP	0.165	0.336	0.239	0.193	0.205	0.313	0.183	0.228
HOG	0.368	0.343	0.504	0.381	0.569	0.464	0.274	0.315
GIST	0.310	0.217	0.469	0.361	0.602	0.340	0.225	0.318
Cosine	0.605	0.652	0.488	0.306	0.652	0.400	0.344	0.322
LD	0.614	0.526	0.588	0.515	0.615	0.414	0.295	0.281

Table 4 F1-scores of experimental result which character basis(k=1)

	joy	fear	surprise	angry	sorrow	none	hate	shame
LBP	0.421	0.406	0.426	0.257	0.399	0.249	0.196	0.272
HOG	0.631	0.504	0.621	0.504	0.571	0.444	0.288	0.320
GIST	0.482	0.477	0.532	0.414	0.489	0.287	0.279	0.303

5. DISCUSSIONS

To validate the effectiveness and the problem of the proposed method, the success and failure examples of the facial expression classification are analyzed by showing the obtained similarities between the emoticons. Table 5 shows the top three similar emoticons obtained when we used the HOG feature by calculating the Levenshtein distance on the character basis. The method outputted "anger" as the emoticon most similar to the emoticon of facial expression of "anger".

Table 5 Success example(HOG + LD)

Input emoticon: (＃▽皿▽)= 3		Facial expression: anger	
Rank	Emoticon	Facial expression	LD
1	(←▼皿▼#)	anger	3.502
2	(;-ω-)=3	hate	3.676
3	(; ・ ∀ ・)= 3	anxiety	4.038

Although the emoticons expressing different facial expressions had been outputted on the second or lower rank, the most similar emoticon had been obtained correctly. It is considered that the features obtained from "▽" and "▼ became similar to each other by using the image feature. The success also might have been caused because a two-byte character "＃" and a single-byte character "\#" had been able to be recognized as similar characters.

6. CONCLUSIONS

This paper proposed a method to classify facial expressions focusing on shape features expressed by the characters forming the emoticons. The proposed method extracts shape image features by converting an emoticon into an image and classifies facial expressions based on similarity between the feature vectors by using the k-Nearest Neighbor Method. The results of the evaluation experiment showed that the proposed method can classify facial expressions more effectively than the baseline method based on the character features.

In the future, we would like to focus on the differences between the symbols that cannot be recognized by image features, and improve the accuracy of the method to classify facial expressions by implementing the algorithm to clarify such differences by providing clustering processing on the characters in advance

ACKNOWLEDGMENTS

This work was supported by JSPS KAKENHI Grant Numbers 15K00425, 15K00309, and 15K16077. And also, this work is partially supported by Tateishi Science and Technology Foundation.

REFERENCES

Suzuki, N, and Tsuda, K(2006). Express emoticons choice method for smooth communication of e-Business, KES 2006, Part II, LNAI 4252, (pp. 296-302).

Urabe, Y, Rzepka, R, and Araki, K(2013). Emoticon recommendation system for effective communication, Proceedings of the 2013 IEEE/ACM International Conference on Advances in Social Networks Analysis and Mining, (pp.1460-1461).

Yamada, T., Tsuchiya, S., Kuroiwa, S. and Ren, F(2007). classification of facemarks using N-gram, 2007 Int. Conf. on Natural Language Processing and Knowledge Engineering, (pp.322-327).

Ptaszynski, M., Maciejewski, J., Dybala, P., Rzepka, R. and Araki, K(2010). CAO: A Fully Automatic Emoticon Analysis System. Proceedings of Twenty-Fourth AAAI Conference on Artificial Intelligence. (pp.1026-1032)

Okumura, N(2016). An Estimation of the Original Form of Kaomoji using Classifier, SIG Technical Reports of Natural Language Processing, Vol. 17, (pp.1-4), (in Japanese)

Ojala. T, Pietikainen. M, and Harwood. D, A(1996). Comparative Study of Texture Measures with Classification Based on Feature Distributions, Pattern Recognition, Vol.19, Issue.1,(pp.51-59).

Olivia, A., and Torralba, A(2001), Modeling the shape of the scene: a holistic representation of the spatial envelope. International Journal of Computer Vision, Vol. 42(3): (pp.145-175).

Dalal. N, and Triggs. B(2005). Histograms of oriented gradients for human detection, IEEE Conference on Computer Vision and pattern Recognition (CVPR), (pp.886-893).

Ekman, P. and Fresen, W, V(1971). Constants across cultures in the face and emotion. Journal of personality and social psychology, Vol.17(2), (pp.124-129).

Kaomoji Café(2016), Accessed via http://kaomoji-cafe.jp/ on 9 Nov 2017.

Kaomoji.ru(2012), Accessed via http://kaomoji.ru/en/ on 9 Nov 2017.

Kaomoji Copipe(2016), Accessed via http://kaomoji-copy.com/ on 9 Nov 2017.

FACEMARK PARTY(1999), Accessed via http://www.facemark.jp/facemark.htm on 9 Nov 2017.

Proc. 1st International Conference on
Machine Learning and Data Engineering (iCMLDE2017)
20-22 Nov 2017, Sydney, Australia

Real-Time Iris Center Detection and Tracking From Video Images

Dipankar Das[1], Md. Golam Rashed[1], Md. Rezaul Bashar[2], and Yoshinori Kuno[3]

[1]University of Rajshahi, Rajshahi, Bangladesh
[2]Senior Research Fellow, Science, Technology and Management Crest, NSW 2566, Australia
[3]Saitama University, Saitama-shi, Saitama, Japan

Corresponding author's E-mail: dipankar@ru.ac.bd or dipankar_ru@yahoo.com

Abstract

In this paper, an efficient real-time iris center detection and tracking system has been proposed based on the vector field of image gradients (VFOID). The system uses video frames from cost-effective low-resolution USB cameras and natural lighting condition instead of high resolution sensors and infrared illumination. The proposed algorithm includes multi-stage approaches. Firstly, the AdaBoost-based cascaded classifiers are used to detect and track face region and its location in the image. Then, the facial feature points in the face region are extracted to locate the rough area of the eyes. Finally, the VFOID are used to accurately detect and track the iris center within the estimated eye areas. The experimental result shows a high accuracy on existing database and in a real-world environment.

Keywords: *Iris Center Detection, Iris Tracking, Image Gradients, Face Detection and Tracking.*

1. INTRODUCTION AND MOTIVATION

Analysis and understanding of human behavior, particularly of head movement and eye gaze behavior, has been a subject of interest for many years in cognitive psychology and neurophysiology (J. S. Agustin, J. P. Hansen & J. Mateo, 2008); however, a full understanding of the causes, dynamics, and control mechanisms of head and eye movements is still a subject of active research (E. G. Freedman, 2008). Recently, researchers in the computer vision community have increasingly sought to incorporate information gleaned from patterns of human behaviors into intelligent human-machine interfaces (A. Doshi & M. M. Trivedi, 2012) (R. Valenti, N. Sebe & T. Gevers, 2012). Iris center estimation is important in many application areas spanning from human-computer interaction (HCI) to human behavior analysis. The visual gaze, determined by detecting and tracking the center of the iris, provides important information about the interest of the subject. It is commonly used in various application areas such as: controlling devices for disabled people (J. S. Agustin, J. P. Hansen & J. Mateo, 2008), to analyze the user attention while driving (A. Doshi & M. M. Trivedi, 2009), HCI.

In this paper, we propose a real-time iris center detection and tracking approach that uses the gradient vectors of an image within an estimated region of eyes. The existing researches on iris center detection and tracking system can be classified into two broad categories: (i) non-intrusive, and (ii) head mounted devices. In non-intrusive approach, the researchers use external cameras to detect the eye center of the target human (E. Murphy-Chutorian & M. Trivedi 2009) (N. Robertson & I. Reid, 2006). However, in this approach the iris center detection task is very difficult if we use only eye or head information for iris because of the variability of the human head orientation. On the other hand, the head mounted device allows detecting and tracking iris center without having the difficulties of the orientation of the subject's head. In this research area, most of the existing system uses infra-red (IR) lighting properties to accurately detect the pupil position in the eyes (Z. Zhu, K. Fujimura & Q. Ji, 2002) (T. Ohno & N. Mukawa, 2004). However, when head mounted devices are difficult to use or IR lighting is impracticable, we can use image based methods to estimate the center of the iris position. In this research, we use a multi-stage image based method to detect and track the center of the iris position. In our approach we first detect and track the face area in the input video images using efficient face detection and tracking technique. Then based on the detected face area, we extract the facial feature points on the face images. Depending on the facial feature point, we estimate the possible eye location and rectangular area around the eye location. Finally, within the estimated eye region, we detect and track the iris center using the VFOID.

2. PROPOSED APPROACH

The proposed iris center detection and tracking system is illustrated in Figure 1. Here, we use a USB camera situated in an environment to record the eye images of the target person. The recorded video image is used to detect and track the iris center. For this purpose, the system first detects and tracks the face in the input video images. From the detected face region, we also calculate the head position in the image. Within the detected face region, the system tries to find the facial feature points. Based on the facial feature points, we estimate the eye rectangular regions. The rectangular eye regions are used to detect and track the iris center.

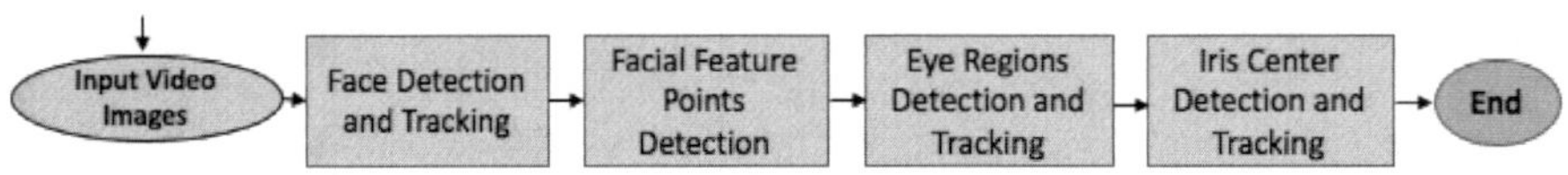

Figure 1. Overview of the proposed iris detection system.

3. FACE DETECTION AND TRACKING

We use 3D head tracking approach proposed by Kobayashi et al. in (Kobayashi, et al., 2006). In the approach, first the AdaBoost-based cascaded classifiers are used for hypothesis evaluation. Then, multiple cascaded classifiers, trained respectively to detect one direction of a human head, are adaptively used in hypothesis evaluation by considering the relationship between human head orientation and camera position. Consequently, tracking a human head robustly and accurately is realized even in the case of observation with low resolution or under varying light conditions.

The AdaBoost-based face detector proposed by Viola and Jones (Viola & Jones, Robust real-time face detection, 2004) has a cascade structure to reduce detection time (Figure 2). This cascade is effective in the evaluation phase even in the particle filter framework. In Figure 2, $\boldsymbol{S}_i$ represents a strong classifier. Each strong classifier classifies an input image into a positive or a negative. Only positive images are used as the input of the next strong classifier. At each stage, a strong classifier is trained to detect almost all face images while rejecting a certain fraction of non-face images. Thus the face detector detects almost all the face images and rejects almost all the non-face images. A strong classifier $\boldsymbol{S}_i(x)$ at each stage of the cascade consists of many weak classifiers $\boldsymbol{s}_t(x)$. This can be described as follows:

$$S_i(x) = \operatorname{sgn}\left(\sum_{t=1}^{T} \alpha_t s_t(x)\right) \tag{1}$$

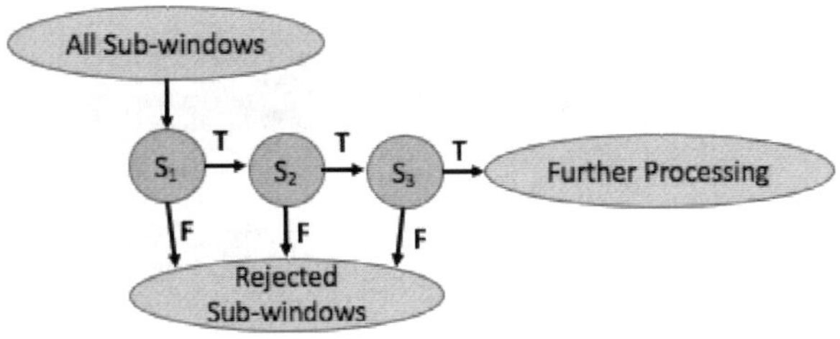

Figure 2. Cascade of classifier.

where T is the number of weak classifiers and $\alpha_t = \log\frac{1-\varepsilon_t}{\varepsilon_t}$. Here ε_t is an error rate specified in the training phase. Each weak classifier $s_t(x)$ evaluates a target image region by using Haar-like features. The detected head position, H_t using the head tracker in the input video frames is shown in Figure 3.

Figure 3. Detected head and its position within a video image.

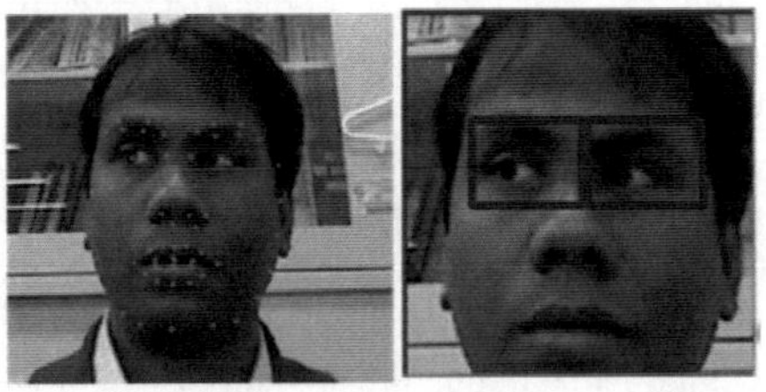

Figure 4. (a) Detected 68 facial feature points, and (b) eye regions detection based on facial feature points.

4. FACIAL FEATURE POINTS DETECTION

The proposed iris detection consists of identifying the pixels that belong to the iris boundary of an image. Because we are in presence of an image representing not only one eye but the full face and even extra scene components, we should consider the detection of face region using the face detection technique as described in the previous section, then the left and right eye regions, before detecting the iris position. To automatically detect the left and right eye regions, we determine the facial feature points. Based on the head location and its area, we detect and track the facial feature points using the active shape model (ASM) (Cootes, Taylor, Cooper, & Graham, 1995) (Figure 4 (a)). The facial feature points are used to roughly estimate the eye regions on the face (Figure 4 (b)).

5. IRIS CENTER DETECTION AND TRACKING

The VFOID provides important information for circular object. The center of a circular object can be detected by analyzing the VFOID as illustrated in Figure 5. We can use same technique for iris center detection if the iris is considered as a circular object. For instant, Kothari et al. (Kothari & Mitchell, 1996) proposed a method that utilizes the flow field characteristics that generates due to the strong contrast between iris and sclera. In their approach, the orientation of each gradient vector is used to draw a line through the whole image and then increases an accumulator bin each time one such line passes through it. The iris center is estimated as the accumulator bin where most of the lines intersect. However, their approach is only defined in the discrete image space and a mathematical formulation is missing. Moreover, they don't consider problems that arise due to eyebrows, eyelids, or glasses. In this research, we use the VFOID within the eye regions to detect the center of the iris within an eye as indicated by the red points in the Figure 6(b).

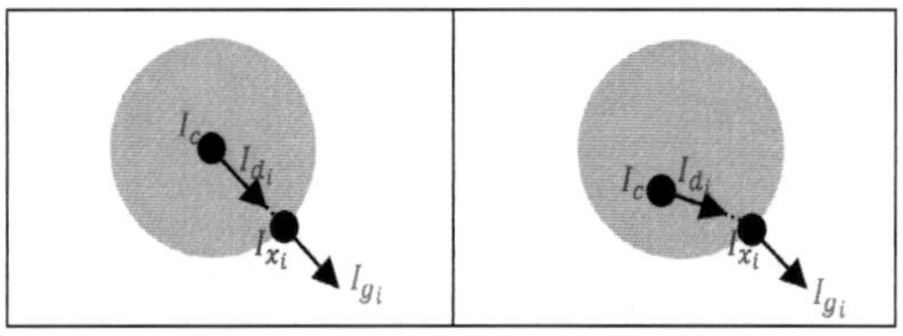
(a) (b)

Figure 5. Abstract view with a dark circle (iris) on a light background (sclera): (a) displacement and the gradient vectors have the same orientations; (b) displacement and the gradient vectors do not have the same orientation.

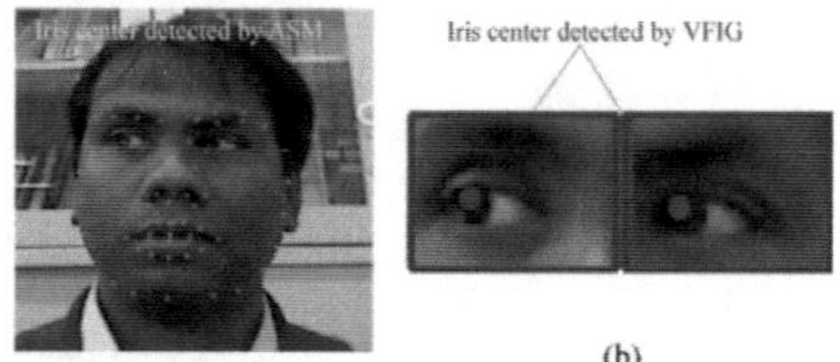

(a) (b)

Figure 6. Iris center detected by (a) ASM landmark points, and (b) VFIG.

Although the facial feature points detected by the ASM model include the eye pupil points, their accuracy is not good for the eye pupil detection as illustrated in Figure 6(a). Thus, we propose the VFIG iris center detection method to accurately detect the iris center in the eye image as follows.

Let I_c be the possible iris center and I_{g_i} the gradient vector at position I_{x_i} . If I_{d_i} is the normalized displacement vector, then it should have the same absolute orientation as the gradient I_{g_i}. We can determine the optimal center I_c^* of the iris (darkest position of the eye) by computing the dot products of I_{d_i} and I_{g_i} and finding the global maximum of the dot product over the eye image. Mathematically,

$$I_c^* = \underset{I_c}{\operatorname{argmax}} \left\{ \frac{1}{N} \sum_{i=1}^{N} (I_{d_i}^T I_{g_i})^2 \right\} \qquad (2)$$

where, $I_{d_i} = \frac{(I_{x_i} - I_c)}{(\|I_{x_i} - I_c\|_2)}$

$i = 1, \cdots\cdots, N$ and $\forall_x: \|I_{g_i}\|_2 = 1$. The displacement vector, I_{d_i} is scaled to unit length in order to obtain an equal weight for all pixel positions in the image.

We also scaled the gradient vector I_{g_i} to unit to tolerate the method against linearity relationship between illumination and contrast of the eye images. To reduce the computational complexity, the proposed method considers only gradient vectors with a significant magnitude, i.e. ignoring gradients vector in homogeneous regions. To obtain the image gradients in an eye image, we compute the partial derivatives using the following equation:

$$I_{g_i} = \left(\frac{\partial I(x_i, y_i)}{\partial x_i}, \frac{\partial I(x_i, y_i)}{\partial y_i} \right)^T \qquad (3)$$

However, other methods for computing image gradients will not change the behavior of the objective function significantly.

6. EXPERIMENTAL RESULTS

6.1. Database

To evaluate the proposed iris center detection technique, we used the BioID face database (BioID face database). It is one of the most challenging databases and many eye center detection researchers use this database to evaluate their system. The database consisted of 1521 face images of 23 different peoples. The faces of the database were in gray-level and were collected from different locations at different times at different poses. In some face images eyes were completely hidden by strong reflection on the glasses. The face image sizes were 286×384.

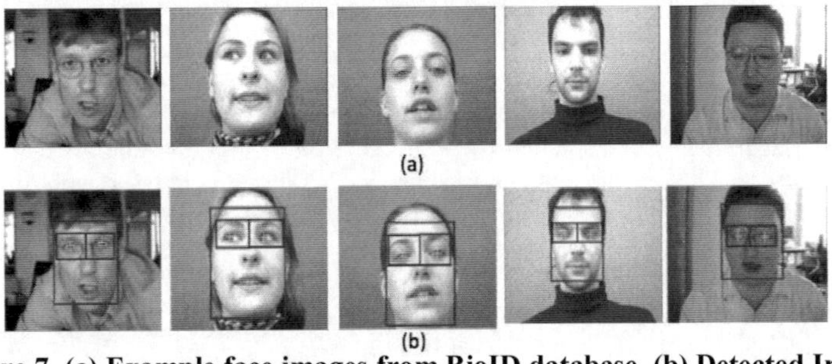

Figure 7. (a) Example face images from BioID database, (b) Detected Iris center.

The database also contained eye position files. The files indicated the *x* and the *y* coordinate of the left eye center (iris center) and the *x* and the *y* coordinate of the right eye center (iris center). Figure 7(a) show some face images from BioID database.

6.2. Evaluation Strategy

We used our face detection technique to detect the face rectangle on the images. Then based on the based on the facial feature points we estimated the eye regions. Finally, the eye region was used to detect the iris center. The accuracy of the iris centers were measured by the normalized error. The technique of this evaluation measure was proposed by Jesorsky et al. (Jesorsky, Kirchberg, & Frischholz, 2001) and is defined as:

$$e \leq \frac{1}{d}\max(e_l, e_r) \quad (4)$$

where e_l = Euclidean distances between estimated and correct left iris center position, e_r = Euclidean distances between estimated and correct right iris center position, and d is the distance between the correct iris's centers.

6.3. Experimental Results and Comparison

Our approach finds the performance for iris center localization which is obtained for $e \leq 0.05$. Figure 7(b) shows some example iris center detection results on the face images from BioID database. The proposed system accurately detected the iris center on the face in different looking conditions. Table 1 shows the performance comparison of our method with other methods for iris center localization. The proposed method performed better than the other three methods presented here.

Table 1: Performance comparison of our method with others.

Methods	Accuracy for $e \leq 0.05$
Asadifard & Shanbezadeh, 2010	47.0%
Kroon, Hanjalic, & Maas, 2008	65.0%
Valenti & Gevers, 2008	77.2%
Our method	***86.1%***

6.4 Real-Time Iris Tracking from Video Data

In real-time analysis of iris center detection, we experimented in a laboratory environment. Here, we placed a webcam in front of the target person's face and s/he was asked to look at different locations in the laboratory. The target person was asked to change her/his position of the head between 0.5 to 2.0 meters from the camera. The person also changed her/his head/eye in left and right directions. Our proposed system detected and tracked the center of the iris position continuously. Figure 8 shows some video frames to illustrate the performance of real-time iris center detection and tracking of our proposed method.

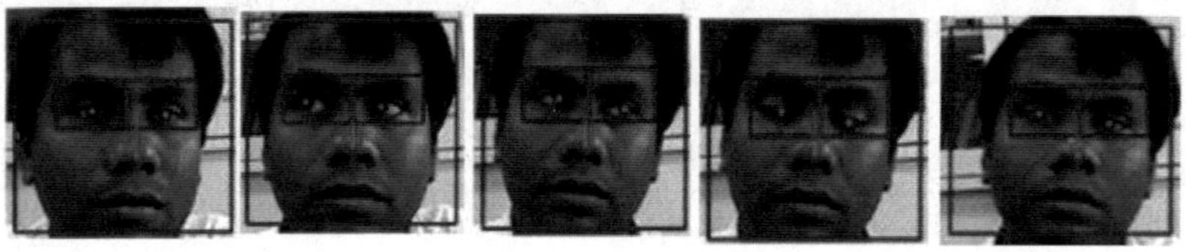

Figure 8. Iris center detection and tracking results on a video images

7. CONCLUSION

In this paper, fully automatic and real-time iris center detection and tracking system is presented. The proposed method can detects and tracks the iris center in various conditions such as background movement, scale changing, and environment illumination changing. The proposed face detection and tracking method effectively detect the face in different posed and scale variations. Then, the system automatically estimates the rough left and right eye regions from the facial feature points. Finally, the VFOID accurately detects the iris center. The proposed system is compared with some state-of-the-art to show the better performance of the system. We also evaluate the proposed method on a real-time tracking of iris center. The future work will focus on the processing of illumination variation, which can cause some highlight reflection on the cornea and increase the difficulty of robust detection of the eye center.

REFERENCES

[1] J. S. Agustin, J. P. Hansen and J. Mateo (2008), "Gaze beats mouse: Hands-free selection by combining gaze and emg," in CHI.

[2] E. G. Freedman (2008), "Coordination of the eyes and head during visual orienting," Experimental Brain Research, vol. 190, pp. 369-387.

[3] A. Doshi and M. M. Trivedi (2012), "Head and eye gaze dynamics during visual attention shifts in complex environments," Journal of Vision, vol. 12, no. 2, pp. 1-16.

[4] R. Valenti, N. Sebe and T. Gevers (2012), "Combining Head Pose and Eye Location Information for Gaze Estimation," IEEE Trans. on Image Proc., vol. 21, no. 2, pp. 802-815.

[5] J. S. Agustin, J. P. Hansen and J. Mateo (2008), "Gaze beats mouse: Hands-free selection by combining gaze and emg," in CHI.

[6] A. Doshi and M. M. Trivedi (2009), "On the roles of eye gaze and head pose in predicting driver's intent to change lanes," IEEE Trans. Intell. Transp. Syst., vol. 10, no. 3, pp. 453-462.

[7] E. Murphy-Chutorian and M. Trivedi (2009), "Head pose estimation in computer vision: A survey," IEEE Trans. Pattern Anal. Mach. Intell., vol. 31, no. 4, pp. 607-626.

[8] N. Robertson and I. Reid (2006), "Estimating gaze direction from low-resolution faces in video," in ECCV.

[9] Z. Zhu, K. Fujimura and Q. Ji (2002), "Real-time eye detection and tracking under various light conditions", The symposium on Eye tracking research & app. (ETRA '02)," NY, USA.

[10] T. Ohno and N. Mukawa (2004), "A free-head, simple calibration, gaze tracking system that enables gaze-based interaction," The symposium on Eye tracking research & app., NY, USA.

[11] D. W. Hansen and J. P. Hansen (2006), "Eye typing with common cameras," in Symp. Eye Track. Res. Appl..

[12] Y. Kobayashi, D. Sugimura, Y. Sato, H. Hirasawa, N. Suzuki, H. Kage and A. Sugimoto (2006), "3D Head Tracking using the Particle Filter with Cascaded Classifiers," in Proc. of the British Machine Vision Conference (BMVC2006).

[13] P. Viola and M. J. Jones (2004), "Robust real-time face detection," International Journal of Computer Vision, vol. 57, no. 2, pp. 137-154.

[14] T. F. Cootes, C. J. Taylor, D. H. Cooper and J. Graham (1995), "Active shape models-their training and application," Computer Vis. and Image Understanding, vol. 61, no. 1, pp. 38-59.

[15] R. Kothari and J. Mitchell (1996), "Detection of eye locations in unconstrained visual images," in IEEE ICIP.

[16] "BioID face database," https://www.bioid.com/downloads/software/bioid-face-database.html.

[17] O. Jesorsky, K. J. Kirchberg and R. W. Frischholz (2001), "Robust Face Detection Using the Hausdorff Distance," in Audio and Video based Person Authentication.

[18] M. Asadifard and J. Shanbezadeh (2010), "Automatic adaptive center of pupil detection using face detection and cdf analysis," in IMECS.

[19] B. Kroon, A. Hanjalic and S. Maas (2008), "Eye localization for face matching: is it always useful and under what conditions?," in CIVR.

Proc. 1st International Conference on Machine Learning and Data Engineering (iCMLDE2017)
20-22 Nov 2017, Sydney, Australia

Development of Gas Accident Prediction Technique in the Area of Dense Energy Consumption by CFD Simulation

Jung-Hoon Kim, Young-Gu Kim and Young-Do Jo
Ph. D., Institute of Gas Safety R&D, Korea Gas Safety Corporation, Eumseong-gun, Korea

Corresponding author's E-mail: jhkim223@kgs.or.kr

Abstract

Accident likelihood is growing due to a correlation for gas and electricity installed in the area of dense energy consumption like traditional market and underground shopping center. In order to prevent accident risks related to gas and electricity in this area, it should be monitored and predicted for factors of gas leak or electricity. Accident risks related to gas and electricity are divided to two cases in the area of dense energy consumption. First case for accident risks is gas leak due to several factors. Then fire and explosion occur from a source of ignitions among many factors. The other case for accident risks is fire caused by several factors. Then gas leak due to damage of gas facilities causes fire spreading and explosion. In this study, gas leak CFD(Computational Fluid Dynamics) simulation was conducted to analyze accident risks related to gas leak. Because of no data for gas density, gas leak simulations were carried out to analyze gas density variation characteristic at gas detector position. Then, the gas accident prediction technique was developed based on variation characteristic of gas leak by using regression and statistic method.

Keywords: Statistic Method, Regression, Computational fluid dynamics, Gas leak Simulation, Accident prevention

1. INTRODUCTION

The areas of dense energy consumption are traditional market and underground shopping center etc. which have a large floating population with facilities of gas and electricity. Accident risks related to gas and electricity are explosion (35.6%), leak (16.8 %) and fire (26.1 %) during the most recent 10 years [Korea Gas Safety Corporation, 2015].

In order to prevent and respond accident risks related to gas and electricity in this area, it should be needed to develop safety management system based on internet of things (IOT) which executes sensor data collection and data analysis for accident prediction and safety control etc. [Tan and Zhang 2013, Zhou et al 2013].

In this study, the technique of gas leak prediction in the area of dense energy consumption was developed for safety management system. To obtain gas density variation data at gas detector position by using computational fluid dynamics (CFD), gas leak simulations were performed by flame acceleration simulator (FLACS) program [GexCon 2013]. The development of prediction technique was established through characteristic analysis for gas density data obtained from CFD simulations. Then, the algorithm of prediction technique was developed for safety management system.

2. ACCIDENT RISK SCENARIO IN THE AREA OF DENSE ENERGY CONSUMPTION

Accident risks related to gas and electricity are divided to two cases in the area of dense energy consumption. First case of A scenario in Figure 1 for accident risks of the area is gas leak due to several factors. Then fire and explosion occur from a source of ignitions among several factors. Areas of traditional markets and underground shopping centers are representative areas of dense energy consumption. In a survey of fire risk factors for the areas during the most recent 9 years (2007-2015), main factors are electrical factors(47.5%), carelessness(24.2%) and unknownness(12.0%) [Disaster Integration Management of Korea 2007-2015]. It recognizes that there is a correlation between gas and electricity. The other case of B scenario in Figure 1 for accident risks is fire caused by several factors. Then gas leak due to damage of gas facilities causes fire spreading and explosion. In order to prevent A scenario of Figure 1, it was studied about a technique of gas leak prediction in this paper.

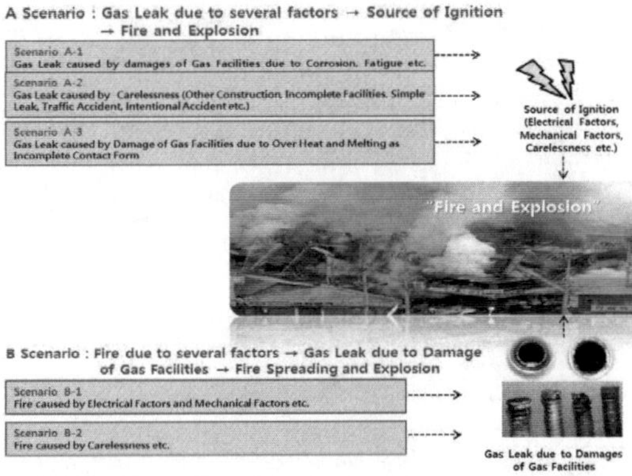

Figure 1. Scenario of accident risk between gas and electricity

Two methods of prediction for gas leak risks are before and after gas leak in the area of dense energy consumption. In the case of before gas leak, deterioration models of wear, corrosion, fatigue and failure rate should be developed for gas facilities. But, it was reviewed that there is serious difficulty for prediction model development due to lack of accident cases and data related to deterioration models. In this study, method and model of risk prediction after gas leak was selected as development method. This prediction model can be obtained to analyze characteristic of gas density data according to time at gas detector position by using CFD analysis. It is needed to suppose trace leakage of the early time point and obtain gas density variation data according to time at gas detector position.

3. GAS LEAK CFD SIMULATION FOR THE AREA OF DENSE ENERGY CONSUMPTION

Because of no data of gas density data about trace leak of the early time point, gas leak simulations were carried out to obtain gas density variation data at gas detector position. Real accidental case was utilized as a place of gas leak simulation. Figure 2 shows target facility of the area of dense energy consumption for gas leak CFD simulation. Accident scene of Figure 2 is gas use restaurant in areas of traditional market. A position of gas leak is defined in the estimated position of accident investigation such as Figure 2. Gas leak automatic shut off device had installed near gas burner. But, this device had been turned off at the time of the accident.

Figure 3 represents photos of target facility for the scene of the accident. Figure 3 (a) is the presumed gas leak position at union part in top of the intermediate valve. Detector of gas leak automatic shut off device was installed in bottom of gas burner such as Figure 3 (b). Figure 3 (c) and (d) shows the photo for the scene of the accident.

Figure 4 is CFD analysis model of target facility about gas leak. To analyze gas leak characteristic in target facility, FLACS program was used [GexCon 2013]. Type of gas in target facility is liquefied petroleum gas(LPG) which is composed propane 95 % and butane 5%. Inner pressure of gas pipe is defined as 25 kPa. Three cases for gas leak flow rate (kg/s) were performed for $10^{-3}\,kg/s$, $10^{-5}\,kg/s$, $10^{-6}\,kg/s$. Analysis cases were determined through several analysis of gas leak flow rate. It was assumed that CFD analysis was modeled in an enclosed space like real accident condition.

Detector of gas leak automatic shut off device should be installed in gas use facilities. Position of detector is a horizontal distance 4 m from center of gas burner in related code (under 0.3 m for LPG) [Korea Gas Safety Corporation 2016]. Figure 4 shows positions of monitor point and leak point. Monitor points are possible positions of detectors. By using CFD analysis for three cases, gas leak density data according to time were calculated at monitor points.

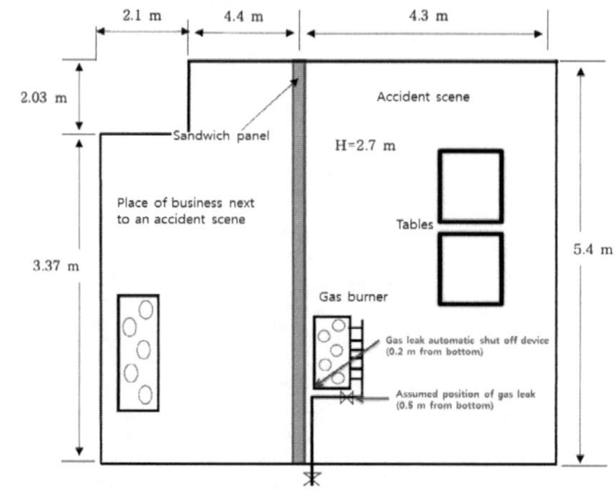

Figure 2. Target facility of the area of dense energy consumption for gas leak CFD simulation

Figure 3. Photos of target facility for the scene of the accident

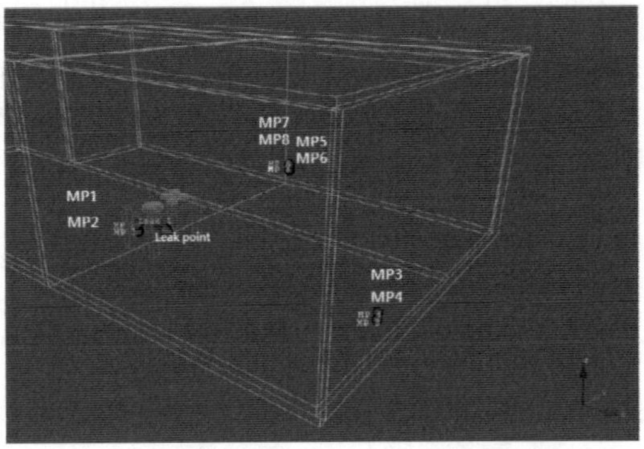

Figure 4. CFD analysis model of target facility about gas leak

4. METHOD OF GAS LEAK PREDICTION TECHNIQUE DEVELOPMENT THROUGH RESULTS OF GAS LEAK CFD SIMULATION

To perform preventive gas safety management, gas concentration and safety action according to 4 steps were proposed for LPG and NG such as Table 1. Existing gas leak automatic shut off device in wired form should be blocked at 25% of lower explosion limit for each LPG/NG. 0.45% of LPG and 1.25% of NG as the first values of 3 step(Warning) in Table 1 are the concentrations of gas shut out. As a part of preventive safety management, two steps (1 step(relief), 2 step(caution)) before 3 step of existing shut out/warning were added to gas safety management step.

Table 1. Step, gas concentration and safety action for safety management of LPG and NG

Division	LPG concentration (Lower explosion limit 1.8%)	NG concentration (Lower explosion limit 1.25%)	Safety action
1 step (Relief)	under 0.05%	under 0.05%	Safety (no action)
2 step (Caution)	0.05~0.45%	0.05~1.25%	Caution
3 step (Warning)	0.45~1.8%	1.25~5%	Warning/Shut off, Dispatch of related organization for gas and electricity
4 step (Danger)	1.8~8.4%	5~15%	Dispatch of related organization for gas and electricity, fire station, rescue center

Figure 5 shows LPG concentration according to time for LPG gas leak flow $10^{-6}\,kg/s$ and $10^{-3}\,kg/s$. It is identified that the relations between LPG concentration and time are nonlinear and linear for each Figure 5 (a) (small flow case) and (b) (big flow case). This relation can be utilized to develop gas leak accident prediction model.

In terms of low concentrations, gas concentration variation according to time was considered such as Figure 6. As LPG gas leak flow increases, variations of LPG concentration according to time increase. This characteristic of concentration data should be applied to accident prediction or gas leak decision.

In wide concentration range, gas concentration variation according to time was reviewed such as Figure 7. The small flow case ($10^{-5}\,kg/s$) and big flow case ($10^{-3}\,kg/s$) of Figure 7 have low and high growth rate respectively for gas concentration. All cases of gas leak flow can be predicted and controlled by data analysis of safety management system. The case of Figure 7 (b) is controlled at concentration 0.45 % by shut off function of detector device due to rapid gas leak.

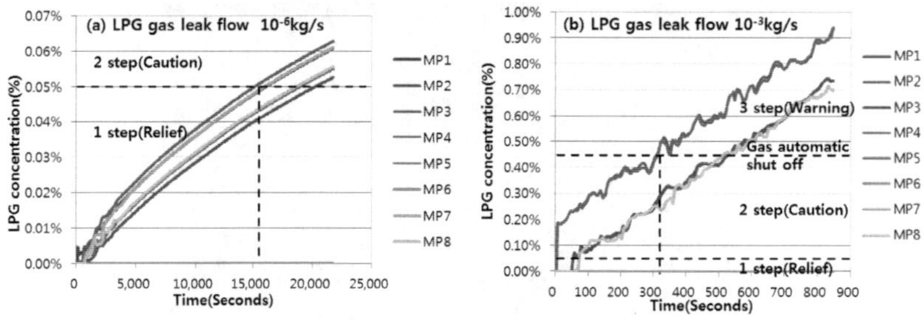

Figure 5. LPG concentration according to time : (a) LPG gas leak flow $10^{-6} kg/s$ **, (b) LPG gas leak flow** $10^{-3} kg/s$

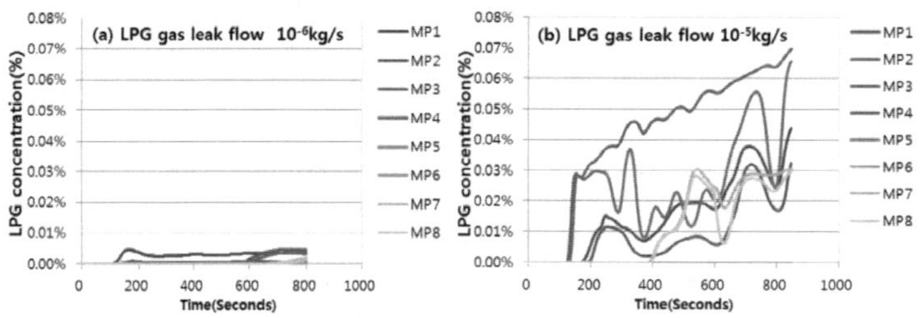

Figure 6. LPG concentration according to time : (a) LPG gas leak flow $10^{-6} kg/s$ **, (b)) LPG gas leak flow** $10^{-5} kg/s$

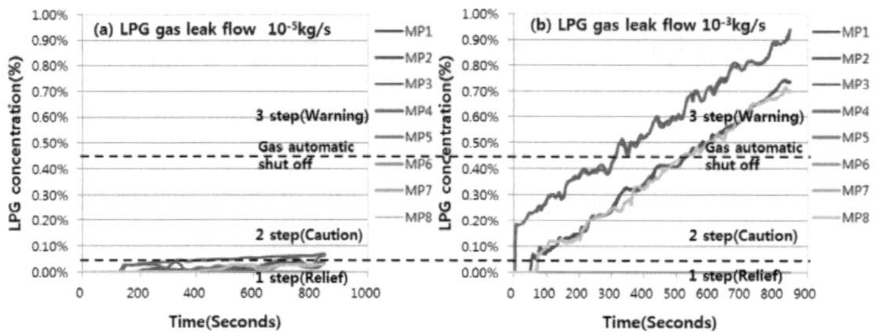

Figure 7. LPG concentration according to time : (a) LPG gas leak flow $10^{-5} kg/s$ **, (b)) LPG gas leak flow** $10^{-3} kg/s$

5. DEVELOPMENT OF GAS ACCIDENT PREDICTION TECHNIQUE CONSIDERING CFD RESULTS

Two methods of prediction for gas leak risks are before and after gas leak in the area of dense energy consumption. In the case of before gas leak, there is serious difficulty for prediction model development due to lack of accident cases and data related to deterioration models. The technique of risk prediction after gas leak was developed considering CFD results. This technique can be applied to safety management system associated with real-time data of gas leak by using wireless sensors.

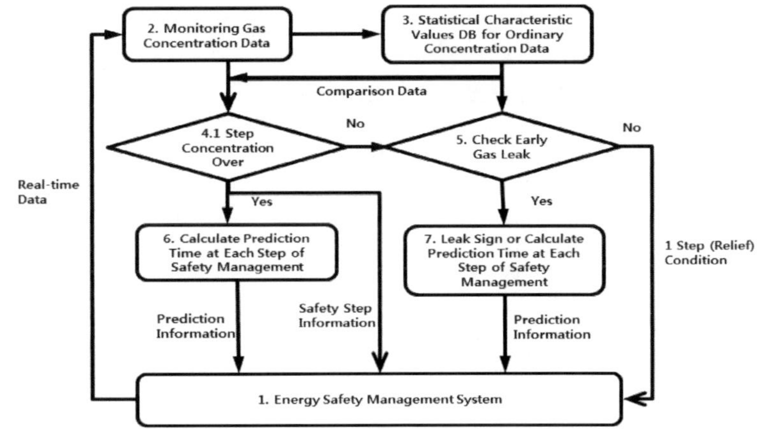

Figure 8. Gas Accident Prediction Algorithm based on Gas Leak Data

The gas accident prediction algorithm was developed considering gas variation data(slope of time-concentration), statistical characteristic variation such as Figure 8.

Energy safety management system is monitoring real-time data of gas concentration by using IoT sensors. Statistical characteristic values for ordinary concentration data as reference data are calculated during certain period. Standard deviation, moving average and gas leak frequency are utilized as statistical characteristic values.

In case of 1 step concentration over (a big rate of leakage) in safety management step of LPG and NG, the safety management system provides predicted information at each safe management step using regression model. In the case of objections (a small rate of leakage), early gas leak is checking through the comparison of statistical characteristic data variation. Also, if prediction model is generated by regression, this system provides predicted information at 1 step.

6. CONCLUSIONS

Accident likelihood is growing due to a correlation for gas and electricity installed in the area of dense energy consumption. To prevent and respond accident risks related to gas and electricity in this area, it should be needed to develop safety management system which executes sensor data collection and data analysis for accident prediction and safety control etc. In this study, the technique of gas leak prediction in the area of dense energy consumption was developed for safety management system. Gas leak simulations were performed to obtain gas leak concentration variation according to time by FLACS program. Main characteristics of gas leak data variation are linear/nonlinear relation and variability over gas leak flow rate which can be applied to accident prediction or gas leak decision. Then, the algorithm of prediction technique was developed for safety management system. The verification test of the algorithm for accident prediction technique will work in the future.

ACKNOWLEDGMENTS

This work was supported by the Korea Institute of Energy Technology Evaluation and Planning (KETEP) and the Ministry of Trade, Industry & Energy (MOTIE) of the Republic of Korea (No. 20152010500080).

REFERENCES

Korea Gas Safety Corporation (2015). Gas Accident Year book.

Z.L. Tan, and C.L Zhang (2013) "Construction of the safety management system for urban underground business district with the application of IOT," Service Systems and Service Management (ICSSSM). Hong Kong, pp. 743-746, July 2013 [10th International Conference on Service Systems and Service Management].

C.A. Zhou, C. Chen, and H. Ren (2013). "Comprehensive Evaluation on Multiple Constraint Elements of City Underground Space Development and Utilization Based on Analytic Hierarchy Process and Fuzzy Comprehensive Evaluation Method," Applied Mechanics and Materials. London, vol. 357-360, pp. 2754-2758, August 2013.

GexCon AS (2013) FLACS v10.0 User's Manual, Accessed 12/08/13; Available from: http://www.flacs.com .

Disaster Integration Management of Korea (2007-2015). Statistical Yearbook of Emergency Management.

Korea Gas Safety Corporation (2016). Standard on Facilities·Technology·Inspection of the Liquefied Petroleum Gas Use Facilities for Containers.

Proc. 1st International Conference on Machine Learning and Data Engineering (iCMLDE2017) 20-22 Nov 2017, Sydney, Australia

Spoken Term Detection Method Using Suffix Array and Bit-Sequence Representation

Satoru Tsuge[1] and Kenji Kita[2]

[1]School of Informatics, Daido University, Nagoya, Japan, and College of Asia and Pacific, Australian National University, Canberra, Australia
[2]Graduate School of Technology, Industrial and Social Sciences, Tokushima University, Tokushima, Japan

Corresponding author's E-mail: tsuge@daido-it.ac.jp

Abstract

In this paper, we describe a spoken term detection (STD) method for a spoken document information retrieval. In general, STD method detects a query term from the spoken documents which are translated from acoustic signal data to text data by the automatic speech recognition system. Because the automatic speech recognition systems are able to output some types of recognition results, we are available for various types of the translated text data for STD. In this paper, we focus on the syllable-based transcriptions and the word-based transcriptions. Because of detecting the query term from a large size of these transcriptions, a rapid STD method is required. Therefore, we have proposed the rapid STD method using a bit-sequence representation and the suffix array. Our method, first, extracts the sub-sequences from the syllable-based transcriptions, and then converts them into the bit-sequence using a hash function. The STD candidates are retrieved using these bit-sequences. Finally, the distance between the query term and these candidates represented as bit-sequences is calculated by using Dynamic Programing (DP) matching. At the same time, our method searches the query term from the word-based transcription using a suffix array method. Then, our method detects the query term by combining these results. In the workshop of NTCIR10, our method has achieved the best performance in STD task. In this workshop, we have submitted the results under the limit conditions of our method. Hence, in this paper, we conduct STD experiments using NTCIR10 SpokenDoc2 Task under the other conditions and evaluate our method. In this experiment, we investigate the STD performances as a function of the number of the candidates of speech recognition and type of candidates of speech recognition. Experimental results show that our method significantly improve the STD method using each transcription. Therefore, we conclude that our method is useful for the STD.

Keywords: Information retrieval, Spoken document retrieval, Spoken term detection, bit-sequence representation, Suffix array

1. INTRODUCTION

With the rapid growth in the amount of information available on the Internet, large collections of full-text documents have become available. In addition, there are not only text data but also many other kinds of media data, such as pictures, movies, music, and speech on the Internet in recent years. Opportunities for retrieving useful information from these multimedia data have increased. Therefore, information retrieval is now becoming one of the most important issues. We focus on retrieval of speech data in the form of "spoken documents" from among these multimedia data. In fact, a spoken document retrieval task has introduced in an NTCIR-9 workshop, which is described in Akiba, T. et al. (2011). One common type of spoken document retrieval system first translates speech data into text documents using an automatic speech recognizer (ASR), and these text documents then serve as target documents. Next, the system retrieves the required information from these target documents using a conventional text information retrieval method. However, it is difficult to adapt text document retrieval methods to spoken document retrieval tasks because there are usually some recognition errors and out-of-vocabulary (OOV) terms in the target documents. To avoid the OOV problem, sub-words such as syllables are used as the index terms for VSM instead of words in Turnen, V. T. (2008). In Iwata K. et al. (2008), a method using a combination of phone-based and word-based recognition results has been proposed for spoken term detection. In addition, it is known that the size of spoken documents, which are retrieval target documents, are huge. Therefore, rapid retrieval methods are demanded.

We have studied about a spoken term detection (STD) method for the spoken documents retrieval. In general, we can obtain some types of transcriptions, which are text documents translated from spoken documents, when we use the multiple ASR system. From these transcriptions, our method uses a word-based transcription and a syllable-based transcription. These transcriptions might be contained the multiple speech recognition candidates, i.e. N-best results of speech recognition. This method first detects the query term from each transcription. Then, the proposed method combines these detected results as final detection results. In this paper, we propose a STD method using a bit-sequence representation for detection of the syllable-based transcription and a STD method based on a suffix array for detecting the query term from the word-based transcription. To evaluate the proposed method, we conducted STD experiments using the SpokenDoc2 tasks of NTCIR-10.

2. SPOKEN TERM DETECTION METHOD

This section describes the proposed spoken term detection (STD) method. Figure 1. shows an overview of the proposed method. The proposed method consists of three parts: a STD method using a bit-sequence representation for detecting a query term from the syllable-based transcription, a STD method based on a suffix array for detecting a query term from the word-based transcript, and a combination method for the final detection results. In the following sections, we describe the details of the proposed method.

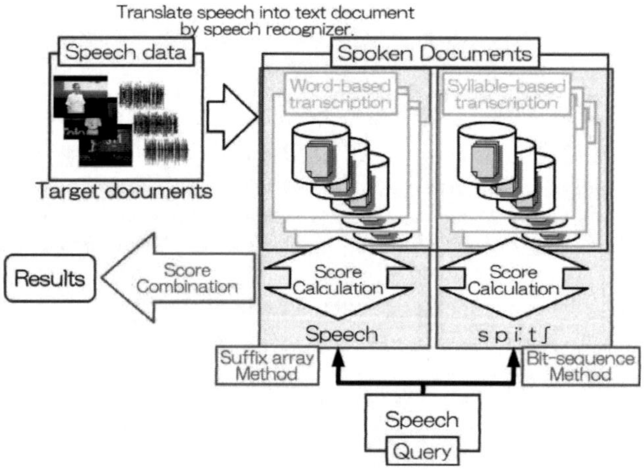

Figure 1. Flow of the proposed method

2.1. STD using bit-sequence representation for syllable-based transcription

This section describes a part of the proposed method, which is the STD method using bit-sequence representation, for detecting the query term from the syllable-based transcription which is results of a sub-word-based ASR. Figure 2. illustrates a flow of the proposed STD method.

Our STD method first extracts the sub-sequences from the syllable-based transcription, which is the target document. Then, these sub-sequences are converted into bit-sequences by using a hash function. Hence, the target documents are represented by a compact bit-sequence. Because the length of the sub-sequence corresponds to the length of the query term, the proposed method needs to construct the multiple bit sequences from the sub-sequences the same length as the query term. Query term is also converted into bit-sequences in the same way. Candidates are detected by calculating a Hamming distance between the bit sequence of the query term and those of the target documents. To represent the target documents in the bit-sequences, only two kinds of bit operations, which are XOR and popcount, are used in this detection process. Therefore, we can obtain the candidates of term detection quickly.

However, this procedure does not use the order of syllables. Hence, after obtaining the candidates of the term detection, the proposed method calculates the edit distances between the query term and the candidates using a DP matching, which is described in Ukkonen, E. (1985), and detects the final results. To calculate the detection score in DP matching, we use the following formula, which is proposed in Katsurada, K., et al (2011).,

$$S = 1/(T/l^{3/2} + 1) \quad (1)$$

where l and T indicate the length of the keyword and the threshold, respectively.

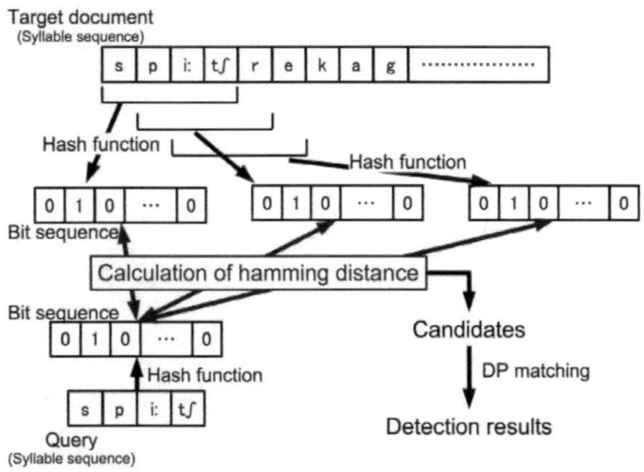

Figure 2. STD method using bit-sequence representation

2.2. STD based on suffix array for word speech recognition system

In this section, we describe a STD method based on suffix array for detection of the word-based transcription. The flow of this method is illustrated in Figure 3.

A suffix array, which is introduced by Manber, U. and Myers, G. (1993), is one of data structures used in full text indices, data compression algorithms, information retrieval, and so on. A suffix array method creates all suffices from a string and these in sorted order. For using this suffix array, this method is low computation cost and low memory space.

In the proposed method, because the target document is the word-based transcription, the word information is known. Because it is supposed that the query term is a word, the suffix of inside of the word can be reduced from the suffix array illustrated in Figure 3. To this end, we can reduce computation cost and memory space.

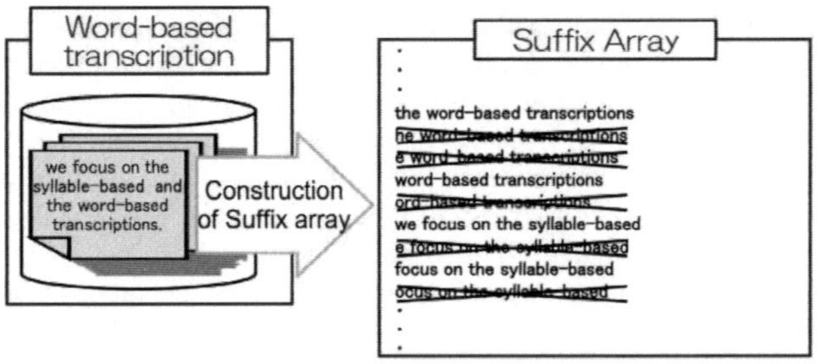

Figure 3. STD method based on suffix array

2.3. Combination method of detection results

In the proposed method, we can get the detection results, which are the results detected by the bit-sequence method from the syllable-based transcription and the results detected by suffix array method from the word-based transcription. The proposed method integrates these results into final detection results. For the integration of these results, the proposed method combines the score of detection results of the syllable-based transcription and the score of detection results of the word-based transcriptions. For the detection score of the syllable-based transcription, we use the DP matching score. The score of the word-based transcription is 1.0 or 0.0 because the search method based on suffix array is the complete-coincidence search method. The proposed method combines these scores for the final term detection result.

3. SPOKEN TERM DETECTION EXPERIMENT

3.1. Experimental conditions

In order to evaluate the proposed method, we conducted a spoken term detection experiment under the condition of the STD large-size task of 2nd round of IR for spoken document (SpokenDoc2) task in NTCIR10 described in Akiba, T. et al. (2013).

The target documents are the spoken lectures in the corpus of spontaneous Japanese (CSJ). The number and the length of the target documents are 2,702 and about 600 hours, respectively. In this experiment, we used the transcriptions of these spoken lectures, which are provided by the NTCIR-10 organizer. These transcriptions are comprised of the word-based speech recognition results and the syllable-based speech recognition results. Two different kinds of language models are used to obtain these transcriptions; one of them is trained by matched lecture text (Matched condition) and the other is by unmatched newspaper articles (Unmatched condition). Each transcription is consisted of the 10-best speech recognition candidates. We used the 100 queries from the formal run of the SpokenDoc2 task at NTCIR-10. The 54 query terms of the all 100 query terms are out-of-vocabulary queries which are not included in the ASR dictionary of the word-based ASR under the Matched condition. We set the threshold of the distance of DP matching to 0.5. We use the top 300 rank detection results for evaluation. The mean average precision (MAP) is used as evaluation measurements.

3.2. Experimental results of each transcription

In this section, we investigate STD performance of each proposed method. Figure 4. shows MAP score of each transcription as a function of the number of the number of speech recognition candidates, i.e., *N*-best speech recognition candidates.

From this figure, we can see that MAP scores using syllable-based transcriptions, which are results of the STD method using bit-sequence representation, are higher than those of word-based transcriptions, which are results of the STD method based on suffix array, regardless of Matched and Unmatched conditions. The STD method based on suffix array can only detect the terms that are completely matched the query term. Hence, the MAP scores of this method are lower than those of the STD method for the syllable-based transcriptions because the suffix array method cannot solve the recognition errors and the out-of-vocabulary words. On the other hand, the STD method using bit-sequence representation retrieves the candidates of term detection then determines the final detection results by using DP matching. Hence, this method can detect the query term even if there are speech recognition errors in target documents or the query term that is out-of-vocabulary word. Therefore, the MAP scores of this method are higher than those of suffix array method.

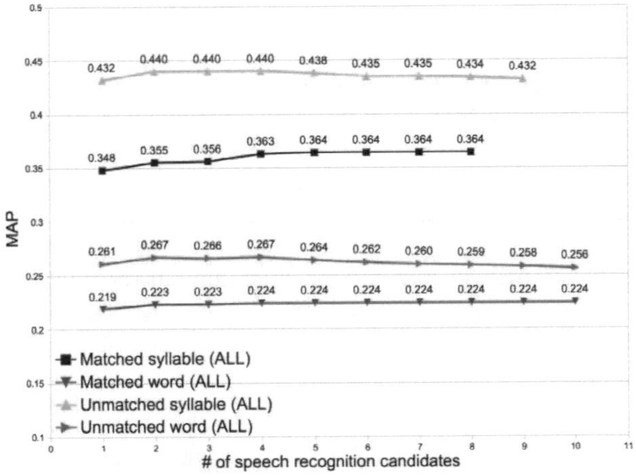

Figure 4. MAP scores of individual methods

3.3. Experimental results of combination method

The proposed method, described in Section **2**, combines the detection results of each transcription, which are the word-based transcription and the syllable-based transcription. Figure 5. shows the MAP scores of the combination results. In this figure, we also show the MAP score, 0.364, under the condition that the number of speech recognition candidates is 5 on Matched condition and the MAP score, 0.440, under the condition that the number of speech recognition candidates is 2 on Unmatched condition using the syllable-based transcription. This figure shows that the combination method improves the detection performance. Actually, the combination method can improve the MAP scores of individual transcriptions, which are 0.224 (word-based transcription) and 0.364 (syllable-based transcription), to 0.388 under Matched condition. In addition, under Unmatched condition, the combination method improves the MAP scores of individual transcriptions, which are 0.267 (word-based transcription) and 0.440 (syllable-based transcription), to 0.468.

However, the MAP score of combination of all transcriptions, 0.443, is lower than that of combination method on Unmatched condition, 0.468. In this experiment, we set the threshold of DP matching to 0.5. When the proposed method combines Matched and Unmatched conditions, the number of syllable recognition candidates is large. Hence, it is considered that the DP matching threshold is not suitable on this combination.

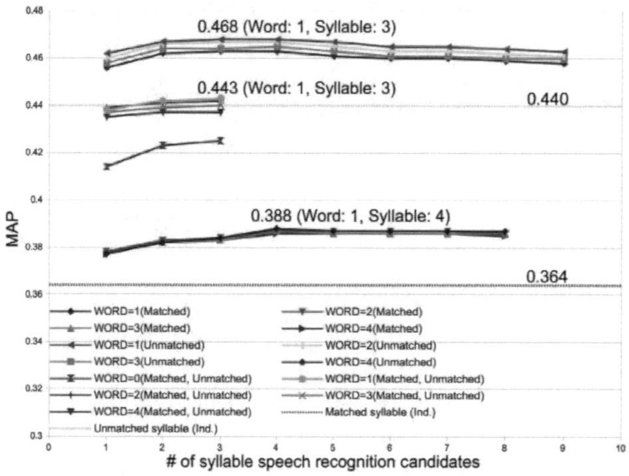

Figure 5. MAP scores of combination method

4. SUMMRY AND FUTURE WORKS

In this paper, we proposed the combination method of multiple detection results for spoken term detection (STD). This method consists of three methods: STD method using bit-sequence for detecting term from the syllable-based speech recognition results, STD method based on suffix array for detecting term from the word-based speech recognition results, and the combination method.

In order to evaluate the proposed method, we conducted the spoken term detection experiments using the large size task in SpokenDoc2 task on NTCIR-10. Experimental results show that the combination of the multiple detection results is able to improve the MAP score of individual method.

In STD method using bit-sequence, we used the sequence search. In future, we will implement the fast method proposed by Indyk, P. and Motwani, R (1998) for reducing the computation cost. We will investigate the details of the experimental results.

REFERENCES

Akiba, T., Nishizaki, H., Aikawa, K., Kawahara, T., and Matsui, T. (2011). Overview of the IR for Spoken Documents Task in NTCIR-9 Workshop, Proceedings of the 9th NTCIR Workshop Meeting, pp. 223–235.

Turnen, V. T. (2008). Reducing the Effect of OOV Query Words by Using Morph-Based Spoken Document Retrieval, Proc. of Interspeech, pp. 2158–2161.

Iwata, K., Shinoda, K., and Furui, S. (2008). Robust Spoken Term Detection Using Combination of Phone-Based and Word-Based Recognition, Proc. of Interspeech, pp. 2195– 2198.

Ukkonen, E. (1985). Finding approximate patterns in strings, Journal of Algorithms, Vol. 6, pp. 132–137.

Katsurada, K., Katsuura, K., Iribe, Y. and Nitta, T. (2011). Utilization of Suffix Array for Quick STD and Its Evaluation on the NTCIR-9 SpokenDoc Task, Proceedings of the 9th NTCIR Workshop Meeting, pp. 271–274.

Manber, U. and Myers, G. (1993). Suffix arrays: A new method for on-line string searches. SIAM Journal on Computing, Vol. 22, No. 5, 935–948.

Akiba, T., Nishizaki, H., Aikawa, K., Hu, X., Itoh, Y., Kawahara T., Nakagawa, S., Nanjo, H., and Yamashita, Y (2013). Overview of the NTCIR-10 SpokenDoc-2 Task, Proceedings of the 10th NTCIR Workshop Meeting.

Indyk, P. and Motwani, R (1998). Approximate Nearest Neighbors: Towards Removing the Curse of Dimensionality, Proc. of 30th Symposium on Theory of Computing, pp. 604–613.

Proc. 1st International Conference on Machine Learning and Data Engineering (iCMLDE2017)
20-22 Nov 2017, Sydney, Australia

Real-Time Panorama Image Stitching based on Multi-Threaded System

Ahyun Lee and Insung Jang

IoT Research Division, Electronics and Telecommunications Research Institute (ETRI), Daejeon, South Korea

Corresponding author's E-mail: ahyun@etri.re.kr

Abstract

This paper proposes a user-friendly panorama image stitching system with a real-time preview. Previous approaches do not have a choice but to the select source images while the user predicts the stitching results through trial and error. Our contribution shows a real-time preview of the stitching results based on the multi-threaded tracking and the blending system. It can help the user easily generate a desired panorama image such as a wide-angle view or building. In our system, the object image designated as an initial frame is tracked to generate a real-time preview. We evaluated the accuracy of the proposed tracking method as compared with a scale-invariant feature transform, such as speeded-up robust features, and oriented FAST, rotated BRIEF. Our experiment results show that our approach can robustly track the object image and provide quality real-time preview images.

Keywords: image-stitching, multi-thread, ORB.

1. INTRODUCTION

Panorama image stitching is a well-known method for creating a wide-angle view image with limited tools or environments. Previous approaches have been concerned with the quality of the panorama image stitching result or fully automated systems as suggested by Cha (2012) and Zhang (2014). And they select the source images, while users predict the stitching results using a guideline for helping capture the source image. To make a view of a user-desired shape, trial and error is required. In this paper, we describe a multi-threaded system based on invariant features for panorama image stitching with a real-time preview. Our contribution is implementing a user-friendly and real-time based stitching system. The predicted stitching result can be shown beforehand in real-time according to the user's controls, as shown in Figure 1(a). This can help the user easily generate a desired panorama image such as a large land-view or building image. The proposed system is composed of detection and blending parts. The blending part is difficult to execute in real time, and thus we configured a real-time executable multi-threaded system. We hope that our multi-threaded approach will be useful to users unaccustomed to such systems.

The rest of this paper is structured as follows. In Section 2, we describe the proposed method in detail in the context of a multi-thread based system. In Section 3, we describe our experiment results. Finally, in Section 4, we provide some concluding remarks regarding our proposal.

(a) (b)

Figure 1. An example of panorama image stitching: (a) real-time preview image and (b) stitching result

2. MULTI-THREAD SYSTEM

Our proposed system is implemented as a multi-threaded system to achieve a real-time performance as suggested by Lee and Hollerer (2009). It can be divided into two threads: the tracking thread tracks the objet image, and the blending thread blends the user-selected source images. The tracking thread is the main thread and is conducted for every frame. The blending thread using an invariant feature based panorama image stitching approach incurs significant computational costs. It can only be applied when the user selects a source image for blending. Prior to the present study, we assume that we calibrate the camera, and its captured image is un-distorted through Zhang's camera calibration as suggested by Zhang (2000).

2.1. Tracking Thread

For the tracking, we use such concepts as an object and scene image. In the initial frame, the user sets the object image for tracking the target. The tracking thread tracks the object image in consecutive frames after the set-up of the initial frame. Our approach uses oriented FAST, rotated BRIEF (ORB) as suggested by Rublee et al (2011) for detecting and matching the features. ORB is a well-known object tracking technique that is rotation invariant and resistant to noise. Moreover, its computational cost is low for real-time applications. However, when using ORB alone, frequent tracking failures occur. In the proposed method, the object features p^{obj} are first detected in an object image using ORB in the initial frame. In addition, we set p^{obj} to object features l^{obj} for the pyramidal Lucas-Kanade tracker (LKT) as suggested by Bouguet (2001). In later frames, l^{obj} is tracked as shown in Table 1.

Table 1. Pseudocode for the tracking thread

FUNCTION: TRACK( )
    Detect object features p^{obj} in the initial frame by ORB
    Set p^{obj} to LKT object features ${l^{obj}}_0$
    $i = 0$;
    BEGINLOOP:
        Track ${l^{obj}}_i$ by LKT, its tracked results are l_i
        Outlier filtering with ${l^{obj}}_i$ and l_i.
        n = number of ${l^{obj}}_i$
        IF: $n < 8$
            Detect scene features p_i by ORB
            Match p^{obj} and p_i, ${m^{obj}}_i$ and m_i is matched features
            FOR: each feature $k \in {m^{obj}}_{i,k}$
                FOR: each feature $j \in {l^{obj}}_{i,j}$
                    IF NOT: check same feature $({l^{obj}}_{i,j}, {m^{obj}}_{i,k})$
                        Set $({m^{obj}}_{i,k}, m_{i,k})$ to $({l^{obj}}_{i,n}, l_{i,n})$
                        $n = n + 1$
        $i = i + 1$
    ENDLOOP

A failed match leads to a preview failure. For robust tracking, we use outlier filtering as suggested by Lee et al (2015) for choosing the good features to track. As a result, if the number of inliers of l_i is lower than eight, which is degree of freedom for a perspective projection transformation, detection of the ORB features is conducted to add more features for robust tracking. The matching results of $m^{obj}{}_i$ and m_i in the *i*-th frame are checked to avoid duplication with features already being tracked. If newly detected features are not tracked, they are added to $l^{obj}{}_i$ and l_i. A warped image is generated as a tracking result in every frame for a real-time preview. To warp the image, a perspective projection matrix **H** is necessary. **H** is calculated based on the relationship between the object features $l^{obj}{}_i$ and the scene image features l_i in the *i*-th frame with a random sample consensus (RANSAC) as suggested by Fischler and Robert (1981).

Tracking reliability depends on the result of computing a perspective projection matrix between an object and the scene image features. Failed tracking causes a failure in computing a perspective projection matrix. In addition, a failed visualization hinders the concentration of the users. In this paper, we use outlier filtering for robust tracking. The proposed outlier filtering classifies the detected features into inliers and outliers based on the relationship of the features between an object image and a scene image as shown Figure 2.

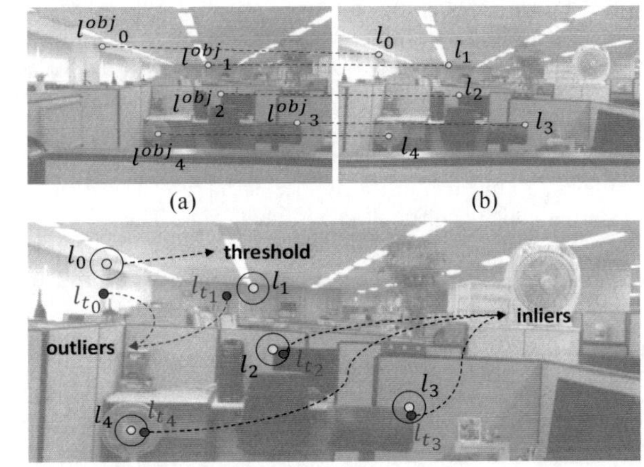

Figure 2. Compute Euclidean distances for outlier filtering: (a) detected features in the object image, (b) tracked features in the scene image, and (c) if a transformed feature in the scene image is out of range, it is an outlier.

Table 2. Outlier filtering for classifying into inliers and outliers

1) Compute a perspective projection matrix **H** from between features $l^{obj}{}_i$ in an object and l_i in the i-th scene image using RANSAC.
2) Compute Euclidean distance $d_{i,j}$ between the j-th feature $l_{i,j}$ and a transformed feature $l_{t_{i,j}} = \mathbf{H} \cdot l^{obj}{}_{i,j}$.
3) If $d_{i,j}$ is smaller than a case-specific threshold, it is an inlier. In other cases, they are outliers.

The main issue is determining the RANSAC distance threshold according to the execution environment for outlier filtering as suggested by Fischler and Robert (1981). An adaptive RANSAC threshold decision enables a case-specific threshold to be determined for each specific frame. In this paper, we update the threshold for every 100 frames.

Table 3. Adaptive RANSAC distance threshold decision

1) Set the initial threshold to $t_0 = s/100$, where the scale s is the length of the diagonal of the camera frame.
2) Euclidean distances $d_{i,j}$ between $l_{i,j}$ and $l_{t_{i,j}}$ are computed until the n-th frame. In the n-th frame, $d_{j_{mean}} = \frac{1}{n}\sum_{i=1}^{n} d_{i,j}$ is calculated from each j-th feature .
3) If $d_{j_{mean}}$ is twice the value of $d_{mean} = \frac{1}{n}\sum_{j=1}^{n} d_{j_{mean}}$, $d_{j_{mean}}$ is removed to exclude impulse errors.
4) Update the case specific threshold t_1 through d_{mean}, the dataset is computed using Otsu threshold decision method as suggested by Otsu (1975).
5) At every n-th frame interval, steps 2) through 4) are repeated.

2.2. Blending Thread

The blending thread begins when the user selects a scene image to be stitched with the object image in the tracking thread. In this work, we use automatic panoramic image stitching with invariant features as suggested by Brown and Lowe (2007). This method uses invariant local features to find matches between all of the images. A probabilistic model verifies the image matches in unordered image sets, and stitches them automatically. Figure 3 shows the procedural flowchart and conceptual time sequence for each process. The first step is to detect and match the scale-invariant feature transform (SIFT) features as suggested by Lowe (2004) in all image sources, which are the selected scene images and the object image. In addition, we find the connected components of the image matches. For each connected component, a bundle adjustment as suggested by Triggs et al (1999) is performed to estimate the camera position, which is difficult to do in real-time. Therefore, we create a new thread for blending. When the blending thread is complete, the stitching result is updated as the object image. ORB features and descriptors are extracted in the new object image for subsequent tracking. Finally, the stitching result is reflected in the real-time preview in the tracking thread.

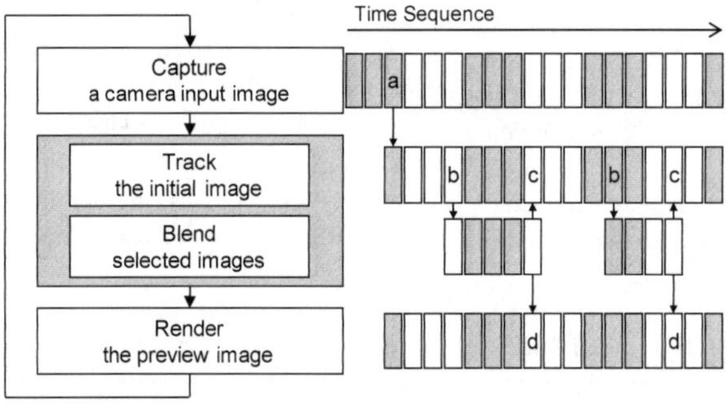

Figure 3. Procedural flowchart and conceptual time sequence for each process. In this figure, a, b, c, and d are as follows: (a) set the initial frame, (b) select the scene image, (c) update the object image, and (d) apply the blend result to the real-time preview image.

(a) **(b)** **(c)**

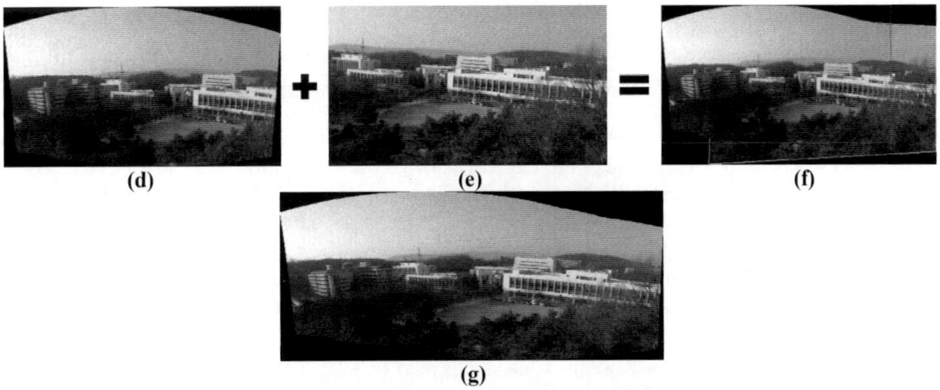

Figure 4. Example of gradual steps for generating a panorama image using our approach with a real-time preview: (a) the first object image, (b) a source image, (c) a real-time preview by (a) and (b), (d) the stitching result of (c) and the second object image, (e) a source image, (f) a real-time preview by (d) and (e), and (g) the final stitching result of (f).

3. EXPERIMENT

The experiments were conducted on a desktop computer with a 4 GHz Intel® core™ i7 CPU using a USB 2.0 camera (Logitech webcam C920). The camera was calibrated using Zhang's calibration method as suggested by Zhang (2000), and its captured images are undistorted to remove the lens distortion. Figure 4 shows an example of the gradual steps used for generating a panorama image by our proposed method with a real-time preview. The real-time preview helps the user generate a user-desired shape. Table 4 shows the process time per frame of each step in our multi-threaded system. The time depends on the camera resolution. For real-time execution with a camera based on 30 fps, the entire process has to be finished within $33ms$. Thus, we use a resolution of 640 × 480 for real-time previews in the tracking thread. The final panorama image will be generated into a 1920 × 1080 resolution, which we already selected for the source images.

Table 4. Comparison of reliability in each tracking approach. The RMSE is the mean value of the x, y, z-axes.

	Time per frame (ms)		
	640×480	1280×720	1920×1080
Tracking thread	**12.4**	**34.71**	**87.18**
Capture & non-distortion	5.07	17.61	37.4
Optical flow tracking	3.8	7.91	13.32
Calculate H	0.82	1.15	3.85
Outlier filtering	0.03	0.04	0.05
Generate preview image	2.68	8.0	32.56
Blending thread	**1662.33**	**2412.4**	**5074.33**
Estimate transform	743.5	1096.0	2467.5
Compose panorama	878.0	1227.4	2449.67
Update the object image	40.83	89.0	157.17

Table 5. Comparison of reliability in each tracking approach. The RMSE is the mean value of the x, y, z-axes.

Index	SIFT	SURF	ORB	Proposed
RMSE (mm)	18.37	39.34	111.39	4.16

In another experiment, we examined the accuracy of the proposed tracking method. We use ORB features for detection and matching. LKT tracks the detected ORB features with outlier filtering in the tracking thread. In our experiments, camera input images were captured while the camera remained fixed. We generate occlusions by hand or through changes in illumination. The root mean square deviation (RMSE) is computed between the camera pose in the first frame without an occlusion and camera poses in the other frames according to SIFT, the speed up robust feature (SURF) as suggested by Bay et al (2006), ORB, and our proposed approach. Because the camera is fixed, the closer the RMSE is to zero, the more stable it is. A camera pose is the fourth column vector of the camera extrinsic matrix using Equation 1. The experiment results illustrate that the proposed method with outlier filtering achieves better results than the other cases, as shown in Table 5.

$$p_i = \begin{bmatrix} f_x & s & c_x \\ 0 & f_y & c_y \\ 0 & 0 & 1 \end{bmatrix} [r_1 \quad r_2 \quad r_3 \quad (x, y, z, 1)^{\mathrm{T}}] \cdot p^{obj} \qquad (1)$$

4. CONCLUSION

This paper presented a user-friendly panorama image stitching approach with a real-time preview. The tracking and blending threads are multi-threaded for a real-time performance. A prediction of the panorama results can be shown through the selection of the scene image in real-time. Therefore, a user can more easily generate a panorama image than in off-line based approaches. To evaluate the accuracy of the proposed tracking method, we compared it with SIFT, SURF, and ORB. The experiment results show that our approach can robustly track an object image and provide quality preview images. We believe that our proposed multi-threaded approach can be useful to users who are unaccustomed to panorama image stitching applications.

ACKNOWLEDGMENTS

This work was supported by MOLIT (Ministry of Land, Infrastructure and Transport), Korea, under the UPA (Urban Planning & Architecture) research support program supervised by KAIA (Korea Agency for Infrastructure Technology Advancement) (grant 13 Urban Planning & Architecture 02).

REFERENCES

Bay H, Tuytelaars T, Gool LV (2006). Speeded-up robust features (SURF), European Conference on Computer Vision, Springer Berlin Heidelberg, 404-417.

Bouguet JY (2001). Pyramidal implementation of the affine Lucas Kanade feature tracker description of the algorithm, Intel Corporation, 5, 1-10.

Brown M, Lowe DG (2007). Automatic panoramic image stitching using invariant features, International journal of computer vision, 74(1), 59-73.

Cha JH, Jeon YS, Moon, YS, Lee SH (2012). Seamless and fast panoramic image stitching, IEEE International Conference on Consumer Electronics (ICCE), 29-30.

Fischler MA, Robert CB (1981). Random sample consensus: a paradigm for model fitting with applications to image analysis and automated cartography, Communications of the ACM, 24(6), 381-395.

Lee T, Hollerer T (2009). Multithreaded hybrid feature tracking for markerless augmented reality, IEEE Transactions on Visualization and Computer Graphics, 15(3), 355-368.

Lee A, Lee JH (2015). Multi-threaded tracker with outlier filtering for spatial augmented reality, International Technical Conference on Circuits Systems, Computers and Communications (ITC-CSCC), 494-495.

Lowe D (2004). Distinctive image features from scale-invariant keypoints, International Journal of Computer Vision, 60(2), 91-110.

Otsu N (1975). A threshold selection method from gray-level histograms, IEEE Trans. Sys. Man. Cybern., 9, 62-66.

Rublee E, Rabaud V, Konolige K (2011). ORB: An efficient alternative to SIFT or SURF, Computer Vision (ICCV), 2011 IEEE International Conference on, 2564-2571.

Triggs B, McLauchlan PF, Hartley RI. Fitzgibbon AW (1999). Bundle adjustment: A modern synthesis, International workshop on vision algorithms, Springer Berlin Heidelberg, 298-372.

Zhang Z (2000). A flexible new technique for camera calibration, IEEE Transactions on Pattern Analysis and Machine Intelligence, 22(11), 1330-1334.

Zhang F, Liu F (2014). Parallax-tolerant image stitching, Proceedings of the IEEE Conference on Computer Vision and Pattern Recognition. 3262-3269.

Proc. 1st International Conference on Machine Learning and Data Engineering (iCMLDE2017) 20-22 Nov 2017, Sydney, Australia

Teleoperated Robot Controlling Interface: an Internet of Things Based Approach

Mohammad Sanaullah Chowdhury[1] and Md. Sharfuddin Parvez[2]

[1]University of Chittagong, Chittagong, Bangladesh
[2]University of Chittagong, Chittagong, Bangladesh

sana1961@gmail.com, sharfuddinparvez00@gmail.com

Abstract

Internet of things (IoT) is an emerging technology that is spreading very fast in today's world. We can see the use of IoT in every sphere of our modern life. It is being said that the next era will be the era of IoT. Robotics is also a very popular research area in this modern world. Robots are like electronic human which has a processor instead of mind and memory instead of brain. As IoT is an Internet based internetwork between machines, it can be used to control a robot. A user-friendly interface can be introduced to build up the interaction between human and robot, through which we can easily control a robot from anywhere over HTTP connection. The idea is to build a teleoperated robot controlling interface using bash language that has low memory consumption and high processing speed. A prototype is being made and discussed in this paper to control a robot using a very cheap IoT core-Raspberry pi with clock speed of 1.2GHz and only using HTML shell scripts, CGI programs and very light HTTP server.

Keywords: Internet of Things, Internet of Robotic Things, IoT core, Teleoperated robot

1. INTRODUCTION

Internet of things (widely known as IoT) is the process of connecting physical devices-where physical devices can be sensor, smartphone, computer and those things which can hold an IP Address. Now-a-days, we use the Internet as via to communicate between human to human. But it is assumed by Tang and Wang (2010) that, the internet will become to the Internet of Things where the communication will be held between machine to machine. We can already see the application of IoT in smart home, smart office, agricultural monitoring, weather forecast, critical infrastructure management etc. An idea about these applications are described broadly by Venkatasubramanian and Gupta (2010), Welbourne and Battle (2009). So why not to communicate with a robot?

A human robot interaction interface can be built by applying the vision of IoT. It will be more effective and easy to communicate with a robot from a remote area by computers or even smartphones. This kind of interaction or internetwork is named Internet of Robotic Things (IoRT). This word is fast introduced by the ABIresearch.

The concept of interconnected devices was introduced by the Carnegie Mellon University in a coke machine at 1982. One of the scientists named Bill Joy realize the necessity of device to device communication as a part of his "Six Webs" framework, which have been retrieved by Pontin (2005) The concept of Internet of Things becomes popular at 1999 when Kevin Ashton (one of the founders of the original Auto- ID center) introduces Radio-frequency identification, described by Magrassi

(2005). He is the one who named the phrase Internet of Things.

As Internet of Things is a new topic in networking, a huge research is done by the researcher to evaluate the machine to machine interaction and reduce human effort in our daily life. We can easily monitor our house ecosystem with our smartphones, open our gate through RFID, monitor our heart rate using heartbeat calculating sensor and smartphone interface and else what. But a small research is done in Internet of Robot things as it is more critical to control a robot through IoT.

Chase (2013) have said in his paper that, there will be 50 billion interconnected devices within 2020 through IoT. So, we can see that not too much work and research is done in the case of robot and IoT. So, there is a lot of scope in this criterion.

The production of single board computers and development of smart embedded systems take the phrase one step further. These boards are used as IoT core. These boards are used in robot technology. These cores are ranging from 1MHz to 3GHz. The most interesting part is that; these cores are very cheap. So, it opens a new door for the smart device developers to experiment their work and make it available to people. Internet of Things and Human Robot Interaction- these two are very popular research topic in today's world. So, if we integrate these two things in one device, it will be more compact and user friendly.

The rest of the paper is organized as follows. The basic concept and applications is discussed in section 2 followed by the system architecture in the next section. The implementation details are described in section 4 followed by the conclusions.

2. CONCEPT AND APPLICATIONS

IoT is an internetwork between devices. So, we can use this concept to build up an interconnection between our smart phones or personal computer's (PC) and a teleoperated robot. Teleoperated robot is a kind of a robot that can be controlled from a remote area. As robotics and IoT are the hotcakes in modern world, it will be interesting to integrate them together and build up a new and sophisticated system. So, we develop an interface through which a teleoperated robot can be controlled. The robot will have a continuous live streaming captured by a webcam. Two things should be taken care of- the interaction time should be small enough and the memory of the processor should be smaller. These things can be achieved by using CGI programs written on bash language and html shell script as scripting language. The prototype can be useful for the developer to work further in the field of Internet of robotic things. The scopes of using the prototype in various aspects are given below:

- The prototype can be used as cheap and very responsive teleoperated robot.
- The prototype can be a cheap surveillance robot which can be used by its master to see what is happening inside and outside of his or her house from a remote area.
- It can be used to monitor poultry farm, crop fields, cattle houses etc.
- It can be used in risky places like mine, militant shelter to monitor the mine or terrorist's activity.

There are many more applications of robot existed in modern world. So, now it is time to make the interaction between human and robot in easier and cheaper way. Hence, everyone can have a personal assistant robot. How about a teleoperated robot with a 1.2 GHz cortex quad core processor and 1 GB of ram within 70\$-80\$? Therefore, we propose the following architecture in next section.

3. SYSTEM ARCHITECTURE

The main goal is to introduce a user-friendly robot controlling system using HTML script and CGI

programs and displaying live streaming using cheap web cam. If we can make an attractive and easy to use interface to control the robot, to monitor an environment that we want to monitor, it will be very useful for us to take a step further to the modern robotics world. If we can control a robot to monitor our home or our children, even the progress of workers in an industry or factory using only the computer or smart phone which we have in front of us in our office or other remote areas, it will be very useful platform for image processing and mapping as the robot will have live streaming video.
The common gateway interface (CGI) executes program that like console applications. It is a way in which user can dynamically interact with the server. A CGI program can be written in any programming language. In this proposed architecture, we use bash language. Bash is a command language. It is first released in 1989 for replacing the Bourne shell in GNU project. It is like a processor that can process commands and can execute command from a file, called a script. Hypertext transfer protocol (HTTP) is a protocol that is used by the World Wide Web. It is an application level protocol that defines the transmission of messages. It is designed to work with request-response messages under client-server communication.

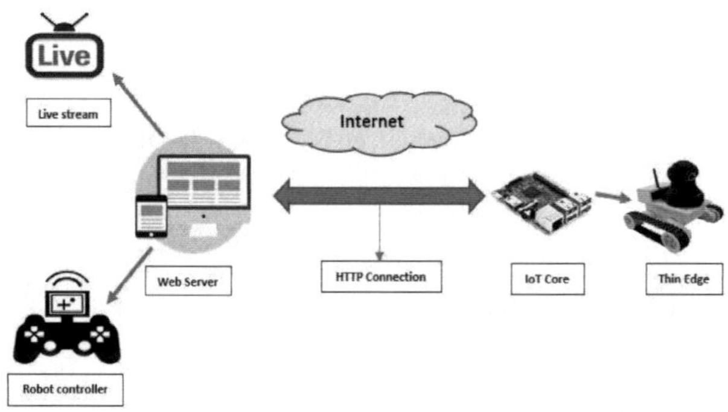

Figure 1. Overall System Architecture

The Figure 1 shows the basic architecture of the proposed prototype. There is an IoT core device mounted on the robot. The core device will work as a HTTP server. A web browser works as the client. There is a webcam connected with the core device. It runs into a specific port and gives live streaming. As depicted in Fig. 1, the thin edge is one of the edge classes that have sensors to collect data. In our case, this is robot framework with a webcam.

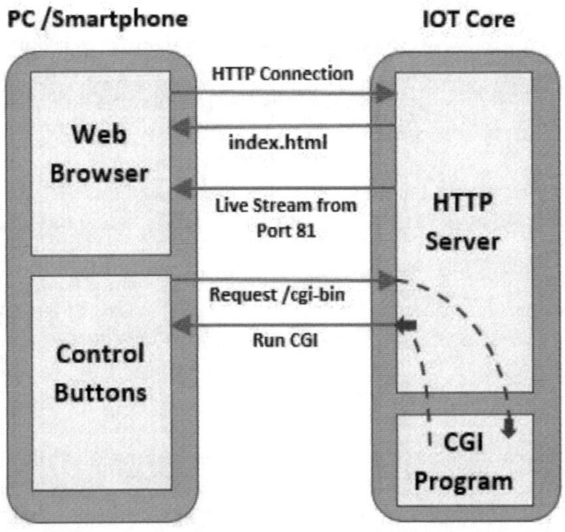

Figure 2. Interaction Between Smart Devices and IoT Core

When the user accesses the IP address of the IoT core connected to the Internet through a browser, a request message is sent to the server and the index.html page is loaded into the browser from the web server. The html page displays the live streaming and there is a controlling interface in the webpage. When a button is pressed, the html shell script requests the correspondent CGI program to the web server. The web server then locates the CGI program and runs it. The CGI program is written on bash language which makes it quicker to send signals to the actuators. The interfaces between the PC/smartphone and IoT core is depicted in Figure 2.

The IoT core is connected to a robotic chassis. It works as the controller board of the robot. The webcam is also connected to the robotic chassis. It gives the live streaming of the forward side of the robot. So, the user can easily navigate the robot.

4. IMPLEMENTATION AND EXPERIMENTAL RESULTS

Under the consideration of proposed architecture, we have implemented following parts:

- We have developed a movement controlling interface for robot using html shell script and CGI programs written in bash language.
- We have used a raspberry pi as the IoT core.
- The raspberry pi works as a webserver with static IP address.
- A webcam is connected with raspberry pi which gives output to a specific port.
- A cheap teleoperated robot is used for image based complex operation with easy and fast acting control interface.

The possible outcome is a GUI which has a controlling interface like a joystick to move the robot forward, backward, left and right in our smart devices. We can easily get access from the remote areas. The method is based on IoT where a raspberry pi is used as one end and our smart device is used at the other end. The prototype consists of a raspberry pi 3 model B as the IoT core, a cheap webcam for capturing videos, 4 DC motors as actuators and L298 motor controller.

The raspberry pi 3 model B is a powerful credit-card size single board computer. It has Broadcom BCM2837 64bit Quad Core Processor powered Single Board Computer running at 1.2GHz and 1 GB of RAM. It has BCM43143 Wi- Fi on board and Bluetooth Low Energy (BLE) on board. There should be a question that why we are using Raspberry Pi as the IoT core? The answer is: it is the cheapest controller that can be used in robotics with a higher clock speed.

To access the gpio's of the raspberry pi, an open source library named WiringPi is used. WiringPi is a pin based gpio access library that is written in C. It can be downloaded and installed without any cost as it is an open source library.

The index.html page consists of two separate blocks- one is for live streaming and the other is for controlling purpose. The live streaming window displays the videos being captured by the webcam at port 81. The controlling blocks consist of five buttons- forward, reverse, left, right and stop. The browser view of the index.html page is given in Figure 3.

There are in total 40 pins in a Raspberry Pi in which 15 can be controlled through request. To assign the behavior of the pins, we have to install the WiringPi library. We have chosen gpio 5, 6, 13 and 19 for giving output to the motor controller. When WiringPi is installed, it will show the state of the gpio's with their corresponding physical address.

There are five CGI programs in the IoT core and they are written in bash language for five operations. Each CGI program defines the state of the gpio pins of the raspberry pi. For example- if we set the gpio 5,6,13 and 19 to OUT the forward movement- all the gpio pins are set to 1. To send request to the web server for CGI programs, html shell scripts are used on button click option.

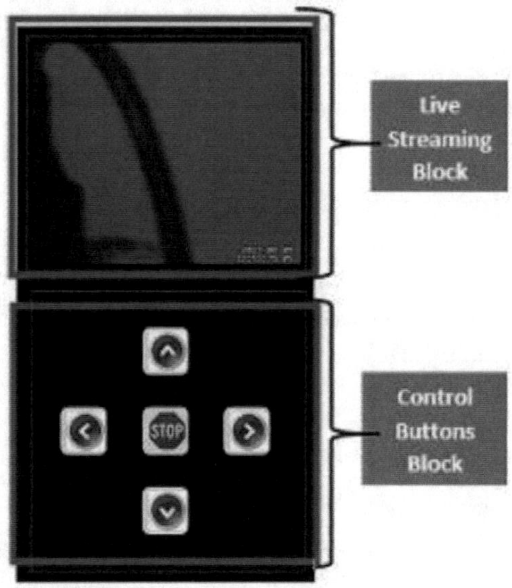

Figure 3. Implementation of Controlling Blocks in the Browser

To have live streaming, a software, named as motion is installed in the raspberry pi. The port number in which the software gives us the streaming is set to 8081. When the motion starts in raspberry pi, it captures video through the web cam and gives output to the assigned port. In the index.html page, this streaming is shown by making the image source to the IP address of the raspberry pi.

Figure 4. The Teleoperated Robot is being Controlled Using Our System

The physical view of our teleoperated robot is given in Figure 4. We have successfully implemented our proposed system and have operated the robot chassis successfully.

The system is developed using shell script and bash language. The same prototype can be developed using Perl script. A comparative result between shell script and Perl script to develop our system is given in Table 1.

Table 1. Comparative result between Shell script and Perl script for our system

Scripting language	**Size with libraries**	**Processor usage for controlling gpio's**
Shell Script	Pre-installed in Linux	5% - 8%
Perl Script	Manually installed (approx. 20 Mb)	13% - 15%

5. CONCLUSION AND FUTURE WORKS

In this paper, Necessities of connecting IoT and robotics are discussed. The paper introduced a prototype robot which can be controlled through HTTP protocol. The reacting time is much reduced over existing systems by using only html shell script and CGI programs. The prototype can be used in near future as a programmable robot and can do many complex tasks with its 1.2GHz processor. The working procedure of the system is described briefly.

It is not so far away when robots will be used in every aspect of our life from morning coffee to late night movie. A standard architecture is presented in the paper which can be used as a reference in this emerging new technology. In future, the prototype can be developed under HTTPS for better security and can be used as a cheap surveillance system. This prototype will take the robot controlling interface based on Internet of Things one step further. Also, the security aspects of the proposed system can be studied as a future work.

REFERENCES

Tan, L. and Wang, N., 2010, August. Future internet: The internet of things. In Advanced Computer Theory and Engineering (ICACTE), 2010 3rd International Conference on (Vol. 5, pp. V5-376). IEEE.).

Venkatasubramanian K. K., Gupta S.K.S., “Interoperable Medical Devices”, IEEE Pulse, September/October 2010, pp 16-27.

Welbourne, E. Battle, et al, “Building the Internet of Things Using RFID: The RFID Ecosystem Experience”,IEEE Internet Computing, Vol. 13, No. 3, 2009, pp. 48-55.

Jason Pontin: ETC: Bill Joy's Six Webs. In: MIT Technology Review, 29 September 2005. Retrieved 17 November 2013.

Magrassi, P. (2 May 2002). "Why a Universal RFID Infrastructure Would Be a Good Thing". Gartner research report G00106518.

Chase, J., 2013. The evolution of the internet of things. Texas Instruments.

Proc. 1st International Conference on
Machine Learning and Data Engineering (iCMLDE2017)
20-22 Nov 2017, Sydney, Australia

Wavelet Based Nonlinear Autoregressive Neural Network to Predict Daily Reservoir Inflow.

W.M.N. Dilini Basnayake[1], Dilhari Attygalle[2], Liwan Liyanage-Hansen[3] and K.D.W. Nandalal[4]
[1]Lecturer, University of Colombo, Colombo, Sri Lanka
[2]Senior Lecturer, University of Colombo, Colombo, Sri Lanka
[3]Senior Lecturer, University of Western Sydney, Campbelltown, Australia
[4]Professor, University of Peradeniya, Peradeniya, Sri Lanka

Corresponding author's E-mail: nadeeka@stat.cmb.ac.lk

Abstract

In spite of the ability of Artificial Neural Network (ANN) to handle nonlinear relationships in data, ANNs fail to predict with high accuracy in the presence of non- stationarity. Hydrological processes in nature exhibits non stationarity due to many interrelated physical and other interrelated factors such as chaotic weather conditions. This paper presents the modelling of one such hydrological process, inflow, to the top most reservoir in the major cascaded reservoir system in Sri Lanka. This daily inflow series has been investigated to be nonlinear and nonstationary. Thus, the difficulties encountered in modelling the inflow series is addressed through a pre-processing strategy based on wavelet transform. Among the methods available in dealing with the nonstationary nature, the wavelet transform was used due to its ability to determine the frequency content of the signal and to assess and determine the temporal variation of this frequency content. The inflow series is decomposed in to several sub series using discrete wavelet transform (DWT). Consequently the appropriate sub series resulted through the wavelet transform are used to model the original inflow using Nonlinear Autoregressive Artificial Neural Network with Exogenous Inputs (NAR-ANN). The results of the NAR-ANN with modified inputs are compared with the results of the base model i.e. NAR-ANN with raw inputs as well as a previously fitted cluster based modular NAR-ANN. The results confirms the superiority of the wavelet based approach over the other approaches, as it has the ability of capturing useful information on various resolution levels.

Keywords: wavelet, pre-processing, nonstationary, NARX, ANN

1. INTRODUCTION

Mahaweli hydropower system is the largest integrated rural development multi-purpose programme based on water resources in Sri Lanka. As such, accurate prediction of the generation of the hydroelectricity from the Mahaweli system is of vital importance to determine the ability of satisfying the future electricity needs and to minimize the wastage of resources. Forecasting the hydroelectric generation, however, requires a comprehensive knowledge of future availability of water resources. Modeling the inflow to the upmost reservoir Kotmale is substantial in that regard, as the entire system depends upon the changes in the natural behavior of this reservoir. A grid search based parameter tuning for the Nonlinear Autoregressive neural network (NAR-NN) was carried out previously to model the daily inflow in a univariate manner (Basnayake et al., 2017 a). It was identified that the fitted NAR-ANN does not adequately forecast inflow and thus be improved. The possible reasons for this less accurate forecast was noted to be the existence of extreme values in the inflow series along with its non-stationarity. It was also proposed through Basnayake et al., (2017 b), to improve the performance of the model incorporating the cluster effect of inflow in order to address the existence of extreme values. This study presents a novel wavelet based application for the same purpose allied with a performance comparison between single NAR-ANN, cluster based NAR-ANN and wavelet based NAR-ANN. Modified codes of MATLAB Neural Network Toolbox and Wavelet Toolbox in Windows platform were used for model fitting.

2. RELATED LITERATURE

This study focuses in forecasting inflow to the upmost reservoir in Mahaweli river basin. It had been identified (Basnayake et al., 2017 a, Basnayake et al., 2017 b) that the inflow variable is nonlinear as well as nonstationary. Moreover the literature (Basnayake et al., 2017 b) reveals that Nonlinear Autoregressive with Exogenous Input (NARX) Artificial Neural Network (ANN), hereafter called as NARX-ANN, was considered a more suitable approach to investigate inflow. Although artificial neural network (ANN) methods have been used extensively as useful tools for prediction of hydrological variables, dealing with nonstationary data has many drawbacks (Cannas et al. 2006; Partal 2009). As one of the remedial measures to this problem, a hybrid modelling approach, namely the Wavelet analysis (WA) combined with ANN is discussed through literature. (Krishna, 2013)

WA is a much used tool for analyzing localized variations of power within a time series. By decomposing a time series into time–frequency space, one is able to determine both the dominant modes of variability and how those modes vary in time. (Torrence and Compo, 1998). Wavelets, due to their inherent properties, have been explored for use in time series analysis. In addition, Wavelet Transform (WT) provides useful decompositions of original time series capturing useful information on various resolution levels. (Krishna, 2013). Continuous type of WT (CWT) determines both the scale contents of a signal and how they vary in time. Discrete wavelet transform (DWT), however, is generally used to decompose a series into sub-signals given proper wavelet and decomposition level. Consequently these sub-signals guide various time series analyses, such as wavelet decomposition, wavelet de-noising, wavelet aided complexity description, and wavelet aided hydrologic forecasting (Smith et al., 1998; Whitcher et al., 2002; Partal, 2009).

There are several applications of DWT in the field of hydrology with the objective of procurement of improved performance. Given below are some of those as an examples that can be extracted from the Literature.

- Rainfall–runoff models [Wang and Lee 1998, Labat et al. 1999, Coulibaly et al. (2000), Cannas et al. (2005) , Rahnama and Noury 2008, Nourani et al. (2011), Okkan (2012a)]
- Streamflow and reservoir inflow forecasting [Smith et al. 1998, Saco and Kumar 2000, Coulibaly et al. (2000), Anctil and Tape (2004), Kisi and Cimen (2011), Adamowski and Sun (2010), Okkan (2012b)]

All studies stated above discuss the idea of improving the performance of nonstationary time series through wavelet transform the efficiency of the DWT approach is confirmed and recommended. It is worth noting that the challenge in wavelet analysis is the selection of the mother wavelet function as well as the decomposition level of signal (Rafiee et al., 2009). Furthermore, the choice of the mother wavelet depends on the structure of the data to be analyzed. Literature had proven that many quantitative methods in finding the best mother wavelet function provide more accurate results than the qualitative methods, but no quantitative method is better than the other. Maximum energy to Shannon entropy ratio criterion [(Yan, 2007), (Kankar et al., 2011)] that is extensively used, is known to captures important information in the series (as combination of some other methods) and is suitable for discrete wavelet transform.

3. MATERIALS AND METHODS

Recall that the inflow series considered in this study is non-linear and non-stationary. Therefore, a wavelet based NAR-ANN approach is considered. Capturing the non-stationarity nature is achieved through DWT, which converts the main series into various sub series in a way that the useful information on various resolution levels are captured.

Given proper mother wavelet $\varphi(t)$ and decomposition level j, the DWT of a series $f(t)$ is operated as (Percival and Walden 2000):

$$w_{f((j,k)} = \int_{-x}^{+x} f(t)\psi_{j,k}^{*}(t)\, dt \text{ with } \psi_{j,k}(t) = a_0^{-j/2}\psi(a_0^{-j}t - b_0 k) \qquad (1)$$

Where a_0 and b_0 are constants, integer k is time translation factor; $\psi^{*}(t)$ is the complex conjugate; $w_j(j,k)$ is the discrete wavelet coefficient under level j. In practice the dyadic (orthogonal) DWT in Eq. (2) is used commonly by assigning $a_0 = 2$ and $b_0 = 1$ (Daubechies 1992):

$$w_f(j,k) = \int_{-x}^{+x} f(t)\, \psi_{j,k} * (t)dt \text{ with } \psi_{j,k}(t) = 2^{-\frac{j}{2}}\psi(2^{-j}\, t - k) \qquad (2)$$

The dyadic DWT allows for the decomposition of a series into a progression of "approximation" and "detail" coefficient sets under each level. In each but except the first stage only approximation coefficients are analyzed (Mallat 1989). Then, sub-signal $f_j(t)$ of original series under certain level $j(j = 1,2, \ldots\ldots, M, M$ is the biggest decomposition level) can be reconstructed by Eq. (3):

$$f_j(t) = \Sigma_k\, w_f(j,k)\psi * (2^{-j}\, t - k) \qquad (3)$$

These sub-signals may correspond to noise or deterministic components, and their sum equals original series. The wavelet used for dyadic DWT must meet the regularity condition with the regularity of order N (Mallat 1989):

$$\int_{-x}^{+x} t^k\, \psi(t)\, dt = 0, k = 1, \ldots, N-1 \qquad (4)$$

In practice seven wavelet families can meet the Eq. (4), namely Haar, Daubechies (db$N$), Coiflets (coif$N$), Symlets (sym$N$), BiorSplines (bior$M.N$), and DMeyer (dmey) respectively (Chui 1992). Besides, the theoretically biggest decomposition level M in dyadic DWT is $log_2(J)$ given the analyzed series has the length of j.

4. ANALYSIS

Wavelet based NAR-ANN was fitted to the daily inflow series obtained from 2013 to 2015 (3 years). Basnayake et al (2017 a) have identified the suitable time period to be 3 years, based on the lowest error in NAR-ANN model fitting through a lookback analysis for the same watershed. The first and the most crucial step was identifying the most appropriate mother wavelet function to carry out the wavelet transformation. As such, 125 mother wavelet functions [Haar, Daubechies (db:2-45), Symlets (sym:2-45), Coiflets (coif:1-5), BiorSplines (bior:1.1, 1.3, 1.5, 2.2, 2.4, 2.6, 2.8, 3.1, 3.3, 3.5, 3.7, 3.9, 4.4, 5.5, 6.8), Reverse BiorSplines (rbior:1.1, 1.3, 1.5, 2.2, 2.4, 2.6, 2.8, 3.1, 3.3, 3.5, 3.7, 3.9, 4.4, 5.5, 6.8) and DMeyer(dmey)] were tested to find out the best suited function for the inflow series of interest.

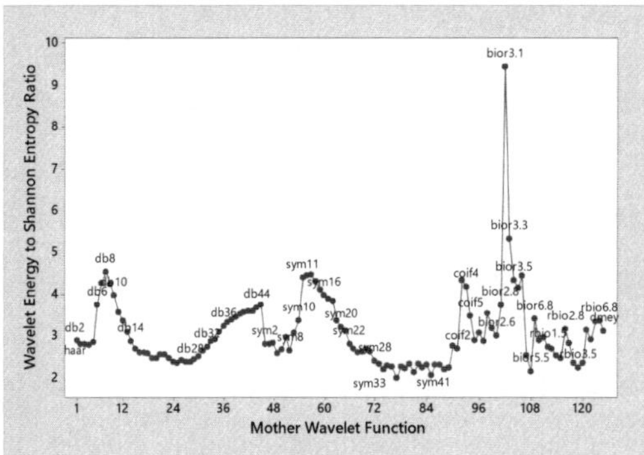

Figure 1: Wavelet energy to Shannon entropy ratios corresponding to DWT of inflow series using 125 mother wavelet functions.

As shown in Figure 2, there is a clear maximum for the "Bio Splines 3.1". The same function was observed as maximum consistently throughout the inflow series for various time intervals, concluding that the most suitable mother wavelet function to transform the inflow is "Bio Splines 3.1".

The main objective of this study is to test the effectiveness in wavelet transform to better capture the non-stationary nature of the inflow series in order to improve the forecast performance. This is achieved by initially transforming the original inflow series into various sub series. Consequently a NAR-ANN model is fitted to each sub-series separately and finally aggregate the results to produce the final output. As explained in the Literature, two key issues in DWT is the mother wavelet selection and the decomposition level choice. The maximum level of decomposition for a particular series is LOG_2 (N) where N being the size of the original series. Thus, for this study the maximum level is 10 i.e LOG_2 (1095).

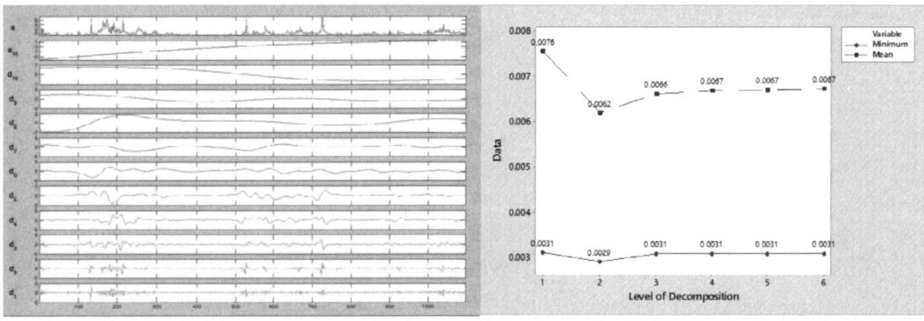

Figure 2: Wavelet decomposed series under maximum decomposition level (left), Minimum and mean grid search errors of wavelet NAR-ANN corresponding to levels 1 – 6.

Since NAR-ANN is a nonlinear modelling structure, fitting linear structures is of no use. As displayed in the Figure 2 (left), nonlinearities associated with the series a10, d7-d10 are very much low. Therefore it was decided to fit NAR-ANN to the wavelet decompositions for levels 1 to 6 only.

The grid search based parameter tuning as stated by Basnayake et al. (2017 a) was carried out in fitting NAR-ANN and the resulted grid errors were analysed in order to identify the best level of decomposition. According to Figure 2 (right), level 2 of decomposition is resulted with both the minimum grid error and the mean grid error. However, after level 4, the errors seems to be consistent with the level of decomposition. Therefore, both levels 2 and 4 were tested in the forecasting, as the results may be different with pointwise errors than with the overall grid search error observed. Figure 3 below displays the graphical view of those two levels of decompositions.

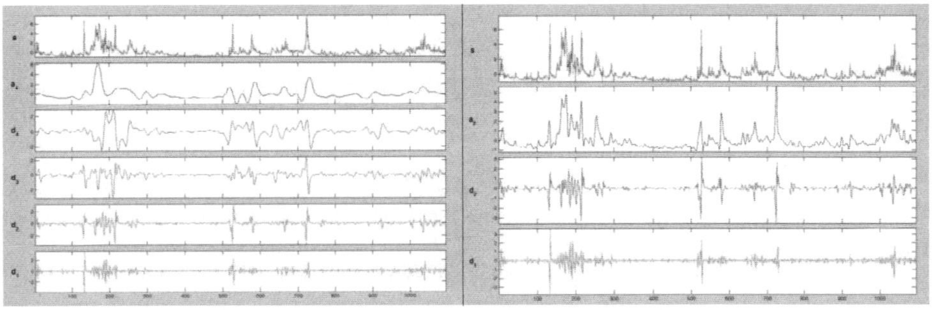

Figure 3: Wavelet decomposed series of level 4 (left) and level 2 (right) [original series (s) in red, approximation in blue and the details in green]

Forecasts for 14 days were obtained for both wavelet based NAR-ANN models and the results were compared with the actual inflow values along with the forecast obtained from single NAR-ANN and cluster based NAR-ANN. The forecasts of the latter two models were obtained according to single NAR-ANN (Basnayake et al., 2017 a) and cluster based NAR-ANN. (Basnayake et al., 2017 b)

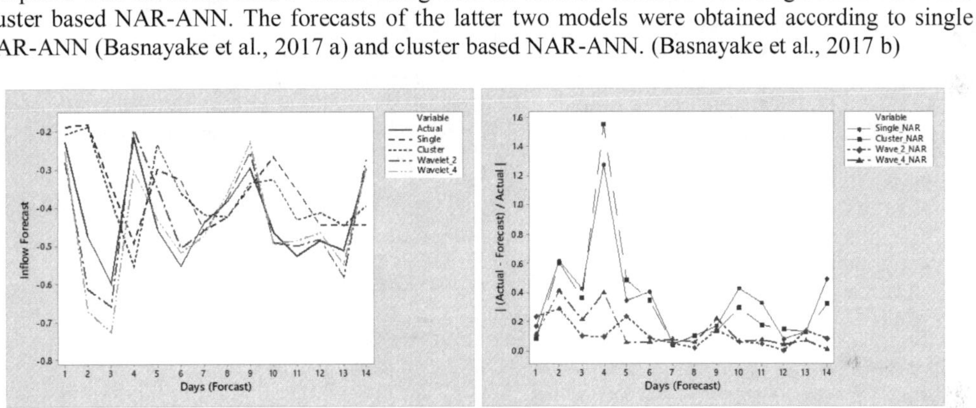

Figure 4: Actual vs. fitted values of 2 weeks ahead forecast of 4 models fitted (left), poinwise errors of the same (right)

As shown in Figure 4, it can be clearly observed that both wavelet based models have outperformed the other approaches at all instances. Pointwise errors with each model were obtained for better comparison of the results. Lowest pointwise errors in general were observed with wavelet NAR-ANN approach with 2 levels. However, there are some instances where the wavelet NAR-ANN with 4 levels produces more accurate results. (With a percentage of 36% where wavelet_4 is better than wavelet_2). The results were analysed in some other parts of the data too and the results were very much similar to above. Even the grid errors as a whole for wavelet based NAR-ANN were lower than that of the other models.

5. CONCLUSION

Wavelet NAR-ANN model has better captured the nonstationary nature of the inflow series when using the approach discrete wavelet decomposition. The divide and conquer approach consisting with 3 decompositions (DWT with 2 levels) has captured the variations in the inflow series very well. Thus the forecast performance of the proposed wavelet NAR-ANN is higher across all the days considered, compared to single NAR-ANN and cluster based NAR-ANN. However, between the two wavelet models compared (3 decompositions and 5 decompositions), around one third (64%) of the time, the model with 3 decompositions has performed better than the one with 5 decompositions. It should be noted that, both models do not show a remarkable difference with respect to the forecast performance. The wavelet NAR-ANN with 2 levels (3 decompositions) is a promising model formulation to obtain

2 weeks ahead inflow forecasts of the Kotmale reservoir in Mahaweli river basin. This paper also illustrates an implementation of the novel approach wavelet NAR-ANN procedure related to daily forecast of a complex non-linear and non-stationary series of data.

6. FUTURE WORK

The favourable performance observed with wavelet NAR-ANN as a univariate approach, can further be investigated by incorporating the exogenous variable rainfall to this approach. Another option for improvement is to adjust for the extreme values and then apply the wavelet NAR-ANN. In addition to this improvements the research can be directed towards applying wavelet based probabilistic forecasting, instead of point forecasts.

7. REFERENCES

Adamowski, J., and Sun, K. (2010). "Development of a coupled wavelet transform and neural network method for flow forecasting of nonperennial rivers in semi-arid watersheds." J. Hydrol., 390(1–2), 85–91.

Anctil, F., and Tape, D. G. (2004). "An exploration of artificial neural network rainfall-runoff forecasting combined with wavelet decomposition." J. Environ. Eng. Sci., 3(1), 121–128.

Basnayake, W.M.N.D., Attygalle, D., & Hansen, L. L., Nandalal, K.D.W. (2017 a). Grid Search based Parameter Tuning of Dynamic Neural Network to Forecast Daily Reservoir Inflow; A Dynamic Neural Network Approach. International Conference Proceedings on Computational Modeling & Simulation (pp. 291-295). Colombo, Sri naLanka.

Basnayake, W.M.N.D., Attygalle, D., & Hansen, L. L., Nandalal, K.D.W. (2017 b). Cluster Based Modular Nonlinear Autoregressive Neural Network to Predict Daily Reservoir Inflow. 17th International Conference Proceedings on Advances in ICT for Emerging Regions (ICTer) (pp. 134-141). Singapore: Global Science & Technology Forum. Colombo, Sri Lanka.

Cannas, B., Fanni, A., Sias, G., Tronei, S., and Zedda, M. K. (2005). "River flow forecasting using neural networks and wavelet analysis." EGU 2005, European Geosciences Union, Vol. 7, Vienna, Austria.

Cannas, B., Fanni, A., See, L., and Sias, G. (2006). "Data preprocessing for river flow forecasting using neural networks: Wavelet transforms and data partitioning." Phys. Chem. Earth, Parts A/B/C, 31(18), 1164–1171.

Coulibaly, P., Anctil, F., Rasmussen, P., and Bobee, B. (2000). "A recurrent neural networks approach using indices of low-frequency climatic variability to forecast regional annual runoff." Hydrol. Processes., 14(15), 2755–2777.

J. A. Schulte, "Wavelet analysis for non-stationary nonlinear time series", Nonlinear Process. Geophys., vol. 23, no. 4, pp. 257-267, 2016.

J. Rafiee and P. W. Tse, Use of autocorrelation of wavelet coefficients for fault diagnosis, Mechanical System and Signal Processing. 23(5)(2009)1554-1572.

Kisi, O., and Cimen, M. (2011). "A wavelet-support vector machine conjunction model for monthly stream flow forecasting." J. Hydrol., 399(1–2), 132–140.

Krishna, B., 2013. Comparison of wavelet based ANN and regression models for reservoir inflow forecasting. J. Hydrol. Eng. http://dx.doi.org/10.1061/ (ASCE) HE.1943-5584.0000892.

Labat, D., Ababou, R., and Mangin, A. (1999). "Wavelet analysis in Karstic hydrology. 2nd Part: rainfall–runoff cross-wavelet analysis." Comptes Rendus de l'Academie des Sciences Series IIA, Earth Planet. Sci., 329(12), 881–887.

Nourani, V., Kisi, O., and Komasi, M. (2011). "Two hybrid artificial intelligence approaches for modeling rainfall-runoff process." J. Hydrol., 402(1–2), 41–59.

Okkan, U. (2012a) "Using wavelet transform to improve generalization capability of feed forward neural networks in monthly runoff prediction." Sci. Res. Essays, 7(17), 1690–1703.

Okkan, U. (2012b) "Wavelet neural network model for reservoir inflow prediction", Sci. Iran., 19(6), 1445–1455.

Partal, T. (2009). "Modelling evapotranspiration using discrete wavelet transform and neural networks." Hydrol. Processes, 23(25), 3545–3555.

P. K. Kankar, Satish C. Sharma, S. P. Harsha, Fault diagnosis of ball bearings using continuous wavelet transform, Applied Soft Computing. 11(2011)2300-2312.

Rahnama, M. B., and Noury, M. (2008). "Developing of Halil River rainfall-runoff model using conjunction of wavelet transform and artificial neural networks." Res. J. Environ. Sci., 2(5), 385–392.

R. Yan, Base wavelet selection criteria for non-stationary vibration analysis in bearing health diagnosis, Electronic Doctoral Dissertations for UMass Amherst, Paper AAI3275786, http://scholarworks.umass.edu/dissertations/AAI3275786, January 1, 2007.

Saco, P., and Kumar, P. (2000). "Coherent modes in multiscale variability of stream flow over the United States." Water Resour. Res. 36(4), 1049–1067.

Sang, Y.F., 2012. A practical guide to discrete wavelet decomposition of hydrologic time series. Water Resour. Manage. 26 (11), 3345–3365.

Smith, L.C., Turcotte, D.L., Isacks, B.L., 1998. Stream flow characterization and feature detection using a discrete wavelet transform. Hydrol. Process. 12, 30–56.

Torrence, C. and Compo, G. P.: A Practical Guide to Wavelet Analysis, Bull. Am. Meteorol. Soc., 79, 61–78, 1998.

Wang, R., and Lee, T. (1998). "A study on the wavelet model of upland watersheds and its application to hydrological estimation." Proc., Statistical Methods for Hydrologic Systems, Vol. 31, 12–21.

Whitcher, B., Byers, S.D., Guttorp, P., Percival, D.B., 2002. Testing for homogeneity of variance in time series: long memory, wavelets and the Nile River.Water Resour. Res. 38 (5),W1054.

Proc. 1st International Conference on Machine Learning and Data Engineering (iCMLDE2017)
20-22 Nov 2017, Sydney, Australia

Dynamic Short Term Load Forecasting using Functional Principal Component Regression

K.A.D. Deshani[1], Dilhari Attygalle[2], Liwan Liyanage-Hansen[3] and G.P. Lakraj[4]

[1]Lecturer, University of Colombo, Colombo, Sri Lanka
[2]Senior Lecturer, University of Colombo, Colombo, Sri Lanka
[3]Senior Lecturer, University of Western Sydney, Campbelltown, Australia
[4]Senior Lecturer, University of Colombo, Colombo, Sri Lanka

Corresponding author's E-mail: deshani@stat.cmb.ac.lk

Abstract

Short term load forecasting plays an important role for the energy industry as accurate predictions are vital in utilizing available resources to optimize the electricity production. Literature reveals that commonly used approaches for such predictions are done using multiple regression, stochastic time series, exponential smoothing, state-space models, neural network models and fuzzy logic, to name a few. In the recent past however, functional data analysis has been an emerging trend to analyse time related data. The electricity demand for a particular day varies with time and as such the daily load curve can be modelled using functional data analysis. Instead of using hourly electricity demand of each day, the dimensionality of the dataset could be reduced using functional principal component analysis. The principal component scores of the selected components were used for prediction. Seasonality and speciality of the day were incorporated using dummy variables and ARIMA models with regressors have been used to predict the principal component scores. Significant variables for the model were identified prior to ARIMA modelling and residual analysis was performed to validate the fitted models. From the predicted principal component scores, the next day's load curve is evaluated. A moving window has been used to make the predictions real time and for the prediction process to be more efficient. This proposed functional analysis based methodology when compared with a commonly used error back-propagated neural network approach have shown improved predictions of the forecasted electricity demands.

Keywords: functional principal analysis, short term load forecasting, time series

1. INTRODUCTION

Electricity demand forecasting or load forecasting is a vital aspect in the energy industry. The prediction can be short term, medium term or long term and each prediction is used at different stages helping the electricity generation companies in their operation and management of the supply to their consumers. Short Term Load Forecasting (STLF) is predicting electricity demand a few hours ahead to one week ahead. The challenge related to STLF is increasing the prediction accuracy. A vast range of literature can be found regarding STLF using multiple regression, stochastic time series, exponential smoothing, state-space models, neural network models and fuzzy logic based approaches indicating the prominence given to STLF. Daily load curve, a plot of electricity demand against time, can be considered as a function with respect to time for which techniques related to functional data analysis can be utilized to do the prediction. In the recent past, functional data analysis has been an emerging trend to analyse time related data due to the promising results obtained through those techniques. The main objective of this study was to forecast the next day's hourly electricity demand using functional data analysis techniques considering the fixed functional patterns of the dataset.

2. LITERATURE REVIEW

A large number of published research regarding short term load forecasting can be found using various approaches. Among them some of the most commonly used technique are multiple regression (Moghram & Rahrnan , 1989), stochastic time series, exponential smoothing (Moghram & Rahrnan , 1989), Bayesian modeling (Cottet & Smith, 2003), neural network models and fuzzy logic (Seetha & Saravanan, 2007, Hsu & Ho, 1992). In the recent past, the use of functional data analysis techniques have shown a rapid increment in various statistical analysis for clustering (Jacques & Preda, 2014), prediction (Antoch, Prchal, De Rosa, & Sarda, 2010, Jornaz, 2016 and Shang, 2013) and system investigation (Gubian, Torreira, & Boves, 2014). Moreover, it can be identified that functional data analysis approaches have been used in various fields such as energy, medical, physics and also environmental sciences. In most of such applications accurate results could be observed using functional data analysis techniques.

When broadly discussing the techniques used in the energy industry for electricity demand prediction, functional data analysis approaches have been used to predict short term (Jornaz, 2016 and Antoch, et al., 2010) and very short term load forecasting (Shang, 2013).In most of the studies the first step is reducing the dimensionality through functional principal component analysis and then the prediction techniques have been used accordingly. The above mentioned studies have compared their research findings with other statistical and artificial intelligence approaches where they have proven that their suggested method using functional data analysis method yield superior results.

3. THEORY AND METHODOLOGY

For this study, the univariate time series of hourly electricity demands was considered as separate load curves representing the electricity demands of each day. In this research, an observation was a load curve for a particular day. The dataset also included the specialty of the day as normal working day, public bank mercantile holiday and New Year's effect. The methodology consists of reducing the dimensionality of the curves using functional principal component analysis and the functional principal component scores of the selected components were used to do the prediction. As the system should be dynamic, new data should be added to the system and the prediction should be performed with the updated data. In order to make the predictions more accurate, a moving window was incorporated to identify the required most recent data that should be used to do the prediction and to exclude past data accordingly. The length of the moving window was also to be chosen and the model validated for the updated data.

3.1. Functional Principal Component Analysis

At a population level, a stochastic process denoted by f can be decomposed into the mean function and the sum of the multiplications of orthogonal functional principal components and uncorrelated principal component scores. It can be expressed as, (Hyndman & Shang, Functional time series forecasting, 2009)

$$f = \mu + \sum_{k=1}^{\infty} \beta_k \phi_k \tag{1}$$

where μ is the unobservable population mean function, β_k is the kth principal component scores, and ϕ_k is the kth population functional principal component. (Shang, Modeling and forecasting for Functional Time Series)

3.2. Forecasting using ARIMA model with regressors

These models combine regression models and ARIMA models to give regression with ARIMA errors. Such models can be used when the errors from a regression are subjected to autocorrelation. (Hyndman & Athanasopoulos, 2012). An ARIMA forecasting equation for a stationary time series is a linear equation in which the predictors consist of lags of the dependent variable and/or lags of the forecast errors. These models aim to describe the autocorrelations in the data. (Hyndman & Athanasopoulos, Forecasting: principles and practice, 2012)

4. DESCRIPTIVE ANALYSIS

For this study, the total hourly electricity demand of Sri Lanka for the period 2008-2012 was obtained from the Ceylon Electricity Board of Sri Lanka (CEB). For model building, data during 2008 - 2011 period was used and data obtained for year 2012 were used for model validation and testing.

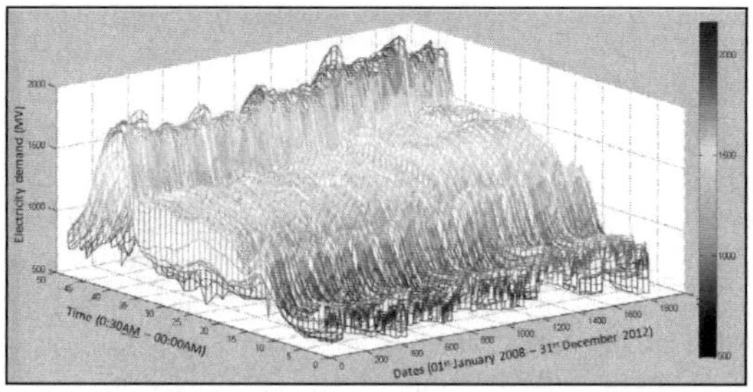

Figure 1: Fluctuations of electricity load curves over the years

When the hourly electricity demands were plotted with respect to each day as in Figure 1, similar shaped daily load curves can be clearly seen. Over the years, the shape seem to be consistent but a gradual increment of the electricity demand can be observed. The electricity consumption within a day varies with respect to time, the day of the week, specialty (whether it a holiday or not), special events and also the weather. These specialties were carefully chosen based on the findings of Deshani, Attygalle, Hansen, & Karunaratne (2014) and Deshani, Hansen, & Attygalle (2014). Figure 2 displays how the load curve varies with respect to day of the week and a holiday. The peak electricity demand is the maximum electricity demand in a particular day usually recorded around 7.00 p.m. In predictions, this peak demand plays a major role as the CEB has to bare a substantial cost due to the use of alternate methods of generating electricity as oppose to using hydroelectricity. (Sri Lanka Electricity Board, 2015)

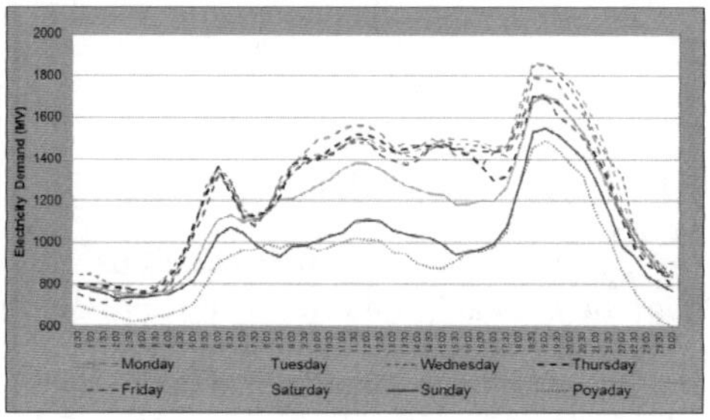

Figure 2: Patterns of daily load curves with respect to specialty of the day

The load curve can be taken as functions with respect to time where they show a fix seasonal pattern among the days as well as years. After a thorough descriptive analysis, different day types that influence the slight changes of the curve shapes were identified.

5. ADVANCED ANALYSIS

This section describes the advanced analysis of this study and it mainly consists of two stages. Dimensionality reduction of the curves using functional principal component analysis and forecasting using ARIMA models with regressors.

5.1. Functional Principal Component Analysis

Firstly, the dimensionality of the daily load curves was reduced using functional principal component analysis. Two principal components could explain nearly 93% of the total variations of the data and their principal component scores were used to do the prediction.

Figure 3 depicts the first two weighted functional principal components and associated scores. The two line plots in the second row of Figure 3 clearly indicate the strong seasonal nature of the first two principal component scores namely Beta 1 and Beta 2.

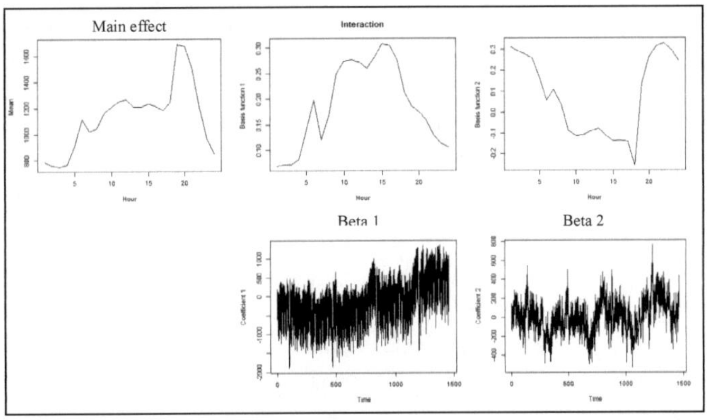

Figure 3: The first two weighted functional principal components and associated scores

5.2. ARIMA Models with Regressors

As there were fixed seasonal patterns, the seasonality could not be just removed by differencing each series. Therefore ARIMA models with regresses were used for the prediction of both Beta1 and Beta2. As regressors, six dummy variables (S1, S2, ..., S6) were added to capture the weekly seasonality of the series. Another four dummy variables (Sp1, Sp2, Sp3, Sp4) were added to incorporate the specialty of the day as 4 categories. Where the five categories represented Category 1: No specialty, Category 2: Poyaday and Public Bank Mercantile holiday, Category 3: Public and Bank holiday, Category 4: Day prior new year and New year day, Category 5: One day after new year and Two days after new year. These categories were previously identified by Deshani et al. (2014) .

Separate regression models were fitted to the two series to check for the significance of the added dummy variables with the index term and the squared index terms. All the regression terms for Beta1 were significant at 5% significance level and for Beta2 all the terms except two variables Sp2 and Sp4 were significant at 5% significance level. Therefore those Sp2 and Sp4 variables were not added as regressors when fitting an ARIMA model for Beta2.

Table 1: Model summary of the fitted regressions for Beta 1 and Beta 2

	Beta 1			Beta 2		
Coefficients	**Estimate**	**t value**	**Pr (>|t|)**	**Estimate**	**t value**	**Pr (>|t|)**
Intercept	-1.056e+03	-36.672	<2e-16	1.514e+02	10.262	< 2e-16 ***
Index	-6.407e-01	-9.119	<2e-16 ***	-2.498e- 01	-6.939	5.94e-12 ***

IndexSq	9.013e-04	19.370	<2e-16 ***	2.407e-04	10.094	<2e-16 ***
S1	1.137e+03	40.954	<2e-16 ***	-2.689e+02	-18.905	< 2e-16 ***
S2	1.228e+03	44.290	<2e-16 ***	-1.827e+02	-12.861	< 2e-16 ***
S3	1.234e+03	44.465	<2e-16 ***	-1.762e+02	-12.391	< 2e-16 ***
S4	1.225e+03	44.166	<2e-16 ***	-1.694e+02	-11.922	< 2e-16 ***
S5	1.179e+03	42.458	<2e-16 ***	-1.639e+02	-11.519	< 2e-16 ***
S6	6.656e+02	24.004	<2e-16 ***	-7.167e+01	-5.044	5.13e-07 ***
Sp1	-1.046e+03	-30.887	<2e-16 ***	1.527e+02	8.869	< 2e-16 ***
Sp2	-2.143e+02	-3.269	0.0011 **	-4.249e+01	-1.265	0.2060
Sp3	-1.573e+03	-15.666	<2e-16 ***	1.023e+02	1.988	0.0471 *
Sp4	-1.542e+03	-15.357	<2e-16 ***	-5.891e+01	-1.145	0.2524
R-Squared	0.8274			0.3463		
F Statistics	578.6 on 12 and 1448 DF			63.91 on 12 and 1448 DF		
p-value	< 2.2e-16			< 2.2e-16		

The residuals of the fitted regression models were used to make the initial guesses for AR and MA terms of the models observing the ACF and PACF plots. Thereafter, the best ARIMA model for each series was selected considering AIC, BIC, and AICc values for several ARIMA models with the regressors mentioned above. Best ARIMA model for Beta1 and Beta 2 was selected based on the lowest AICc, AIC and BIC values and the Ljung Box test. ACF and PACF plots also suggested that the models were adequate for the dataset. The best model for Beta1 was ARIMA(3,1,1) and ARIMA(1,1,3) for Beta2.

5.3. Predicting Next Day's Load Curve

From the predicted functional principal component scores, the electricity hourly demand is predicted using the function $f = \mu + \sum_{k=1}^{2} \beta_k \phi_k$ according to in Equation 1.

5.4. Dynamic Prediction

In order to predict in a dynamic manner, two main aspects were considered. First, it is identified whether the best ARIMA model is consistent with different time periods and then the best size of the moving window is selected. The four years that used to build the model, therefore, were further examined separately as 1 year, 2 year, 3 year and 4 year periods. For each period the best ARIMA models for Beta1 and Beta2 were consistent over the considered period. Therefore for the dynamic model, ARIMA(3,1,1) model was used to predict the first principal component scores and ARIMA(1,1,3) was used to predict the second principal component scores.

Thereafter the most appropriate size of the moving window was is selected. The accuracy and the execution time was considered for selecting the most suitable size of the moving window. A moving window of 1 year time period was used to carry out dynamic forecasting because of the accuracy and less execution time compared to longer window sizes. As such, after each day, a new load curve will be added to the database and the moving window will be shifted by one day to do the next prediction.

6. RESULTS

The test data consisted of daily load curves of year 2012 and the R software was used to do the analysis. **Error! Reference source not found.** shows predictions made for randomly selected two days using the suggested approach. Right side figure includes the predicted daily load curves using one PC and a neural network. The results using the proposed approach seem promising. Moreover, the results were compared with an error back propagated neural network where the features have been selected based on mutual information criteria. The proposed method predicted more accurately than the neural network approach. While comparing the predictions among the different methods, it can be clearly seen that two component predictions yield higher accuracies and can predict the peak time quite accurately than other techniques. Temperature also needed to be added to the model as future work as literature reveals that temperature has a considerable influence on the load forecasting. (Cottet & Smith, 2003, Taylor & Buizza, 2003)

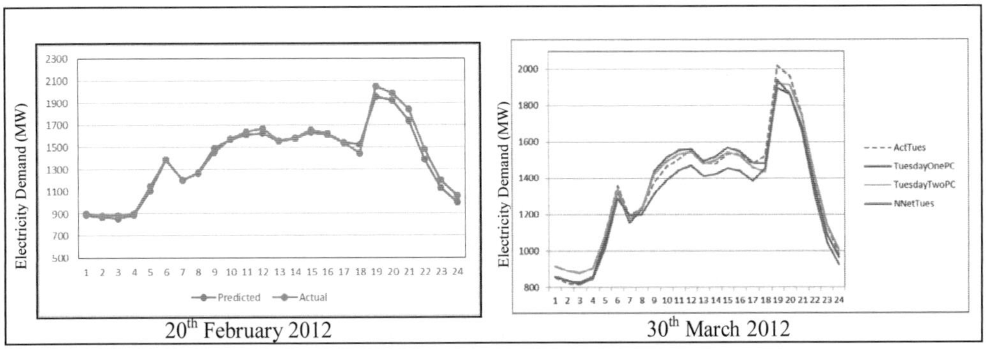

Figure 4: Some predicted daily load curves with actual curves

7. CONCLUSION

Using functional time series regression approach, high accuracies could be obtained for this dataset except for a few situations. The days having less accuracies could be due to extremes weather conditions and by incorporating weather information the unexpected fluctuations of the daily load curves may be captured. In most situations the suggested methodology yielded higher accuracies than iterative error back propagated neural network. Using two component functional principal component the peak time was well captured than other techniques mentioned in this paper.

ACKNOWLEDGMENTS

We acknowledge University of Colombo for the financial assistance provided.

8. REFERENCES

Antoch, J., Prchal, L., De Rosa, M. R., & Sarda, P. (2010, December). Electricity consumption prediction with functional linera regression using spline estimators. Journal of Applied Statistics, 2027-2041.

Cottet, R., & Smith, M. (2003). Bayesian modeling and forecasting of interaday electricity load. Journal of the American StatisticalAssociation, 839-850.

Deshani, K., Attygalle, M., Hansen, L., & Karunaratne, A. (2014). An exploratory analysis on half-hourly electricity load patterns leading to higher performances in neural network predictions. International Journal of Artificial Intelligence & Applications, 5(3), 37-51.

Deshani, K., Hansen, L. L., & Attygalle, M. (2014). Improved neural network prediction performanceof electricity demand:. Second International Conference on Computational Science and Engineering, (pp. 137-147).

Gubian, M., Torreira, F., & Boves, L. (2014). Using Functional data analysis for investigating multidimensional dynamic phonetic contrasts. Journal of Phonetics.

Hsu, Y. Y., & Ho, K. L. (1992). Fuzzy expert systems: an applicatio to short term load forecasting. IEEE proceedings-C, (pp. 471-477).

Hyndman, R. J., & Athanasopoulos, G. (2012). Forecasting: principles and practice.

Hyndman, R. J., & Shang, H. L. (2009). Functional time series forecasting. Journal of the Korean Statistical Society, 199-211.

Jacques, J., & Preda, C. (2014). Functional data clustering: a survey. Adv Data Anal Classif, 231-255.

Jornaz, A. S. (2016). Modeling daily electricity load curve using cubic spline and functional principal components. Scholars' Mine: Doctoral Dissertation.

Moghram , I., & Rahrnan , S. (1989). Analysis and evaluation of five short-term load forecasting techniques. IEEE Transactions on Power Systems,, 4(4), 1484-1491.

Seetha, H., & Saravanan, R. (2007). Short term electricity load prediction using fuzzy BP. Journal of Computing and Information Technology, 267-280.

Shang, H. L. (2013). Functional time series approach for forecasting very short-term electricity demand. Journal of Applied Statistics, 40(1), 152-168 .

Shang, H. L. (n.d.). Modeling and forecasting for Functional Time Series.

Sri Lanka Electricity Board. (2015). Know about Frequently ask questions. Retrieved August 21, 2017, from Ceylon Electricity Board: http://www.ceb.lk/knowledge-center/#tab-1439798656115-4-5

Taylor, J. W., & Buizza, R. (2003). Using weather ensemble predictions in electricity demand forecasting. International Journal of Forecasting, 57-70.

Proc. 1st International Conference on Machine Learning and Data Engineering (iCMLDE2017)
20-22 Nov 2017, Sydney, Australia

Domain Knowledge Driven Approach in Data Analytics

Koh May Fern[1], Dr. Chew Yew Choong[2] and Seck Thiam Ho[3]
[1]Staff Engineer, MIMOS Berhad, Kuala Lumpur, Malaysia
[2] Principal Data Scientist, MIMOS Berhad, Kuala Lumpur, Malaysia
[3] Senior Staff Engineer, MIMOS Berhad, Kuala Lumpur, Malaysia

Corresponding author's E-mail: kohmf@mimos.my, yc.chew@mimos.my, th.seck@mimos.my

Abstract

This paper presents the methodologies and systematic approaches in software design development process which incorporate semantic technology and intelligent knowledge (which also known as "elements" in this paper) identification in Data Analytics. Domain Knowledge (Ontology) approach was applied where users' requirements and knowledge were identified in early design stage. Requirement analysis were also conducted in earlier stage to understand the data model and knowledge management of existing Data Analytics in the market, at the same time finding gaps for continuous improvement in enhancing the quality of the Domain Knowledge. The development and discovery of the Domain Knowledge as well as the process of how to extract useful knowledge for Data Analytics creation were also discussed in the paper. Unlike the conventional process checklist, the Relevancy factor(s) is taking into consideration during the identification of closest matching knowledge which is significant in both Design and Development stage. Relevancy factor(s) is the key prevention method to eliminate irrelevant data and at the same time retain important information for Data Analytics in the later stage. Furthermore, case studies which captured in this paper demonstrated the effectiveness in both research and development of Domain Knowledge in addressing Data Analytics challenging problems.

Keywords: knowledge management, domain knowledge and data analytics.

1. INTRODUCTION

Software development life cycle (SDLC) is systematic method includes a series of stages for organization to implement a product. According to Dennis, Wixom, and Tegarden (2009) "the systems development life cycle (SDLC) is the process of understanding how an information system can support business needs by designing a system, building it, and delivering it to users" [1].

1.1. Conventional System Development Life Cycle (SDLC)

There are 5 major stages in a system development life cycle namely:

- Stage 1 – Requirement Analysis, where new system requirement, user requirement and project scope is defined.
- Stage 2 – Design, where technical architecture and system model is defined.
- Stage 3 – Development, coding and database is construct.
- Stage 4 – Testing, where test script is prepared and execute testing on the system.
- Stage 5 – Deployment, training is conducted prior system implementation.

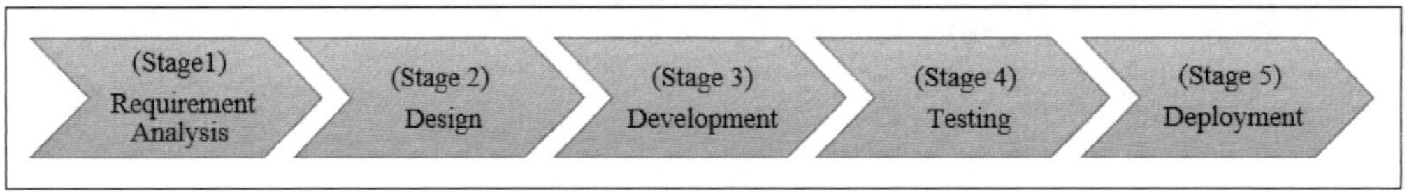

Figure 1. Conventional SDLC in Stages

Figure 1 above illustrated the conventional SDLC practiced in the organization. The process is primarily sequential where stage 2 only start upon completion of stage 1. Therefore, this type of conventional system life cycle often known as process-driven basis.

2. PROCESS OF KNOWLEDGE DISCOVERY AND ENRICHMENT

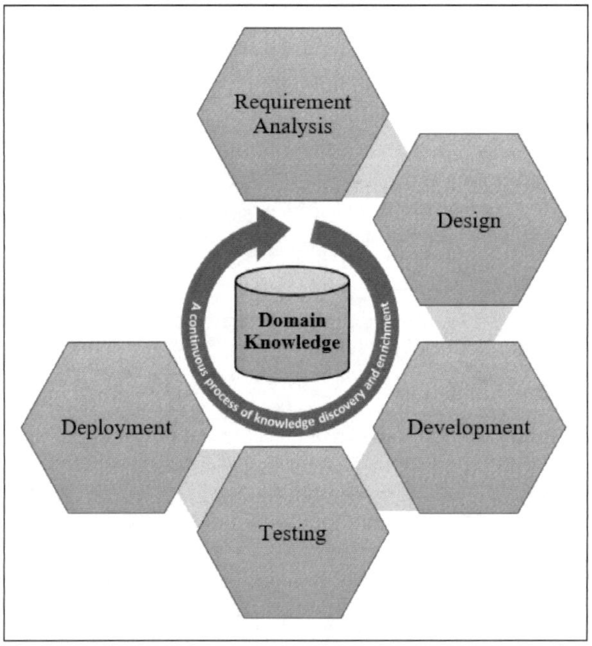

Figure 2. A Continuous Process for Knowledge Discovery and Enrichment

Figure 2 above illustrated the proposed process for knowledge discovery and enrichment in this paper. A continuous process for knowledge discovery and enrichment in which the process is incremental and interactive instead of sequential. As an example, testing can evolves from the beginning of the development, cycle times shortened, with everything trending towards the continuous such as integration, deployment, implementation, and so on. Therefore, this type of process is focused more on result-oriented [2].

3. INTELLIGENT KNOWLEDGE

The intelligent knowledge comprises of all the domain knowledge pertaining to information related to all the processes in software development. The intelligent knowledge is referred as an Ontology which captured the information as well as knowledge of one or many relationships of each of the instances that representing a task or an item in SDLC.

3.1. Ontology

Ontology is a formal naming and definition of the types, properties, and interrelationships of the entities that really or fundamentally exist for a particular domain of discourse. It is thus a practical application of philosophical Ontology, with a taxonomy [3]. In 1993, Gruber originally defined the notion of an Ontology as an "explicit specification of a conceptualization" [4]. Domain ontology can be used as domain knowledge. Figure 3 below illustrated an example of domain ontology for SDLC.

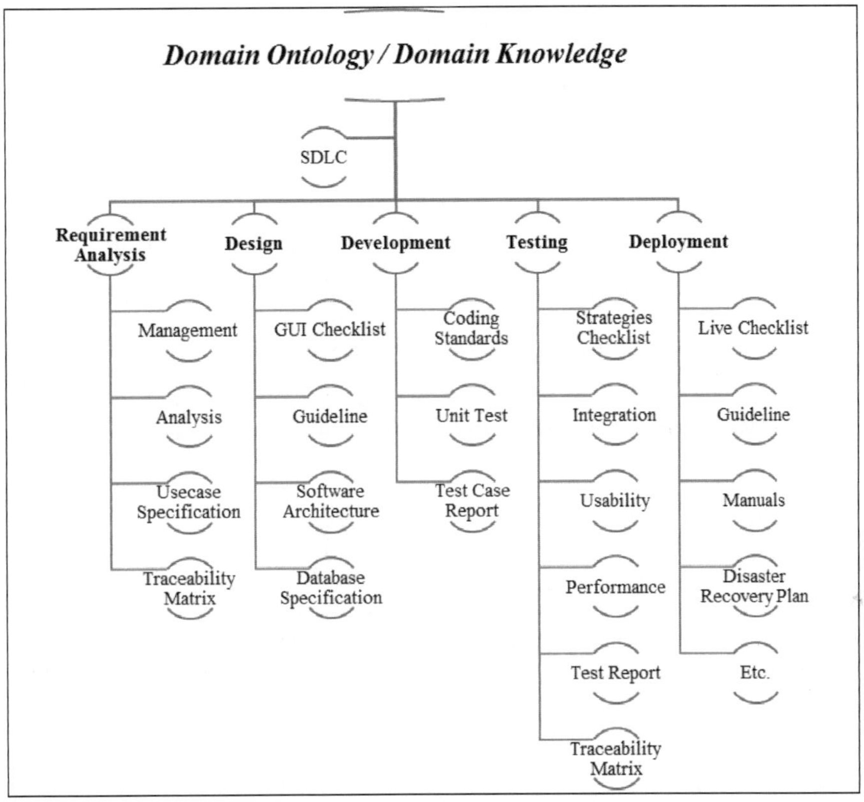

Figure 3. Domain Ontology for SDLC

3.2. Elements

Elements are the key factors that contribute to the level of comprehensive of an Ontology. An elements is references as RDF Schema (Resource Description Framework Schema, variously abbreviated as RDFS, RDF(S), RDF-S, or RDF/S) which is a set of classes with certain properties using the RDF extensible knowledge representation data model, providing basic elements for the description of ontologies, otherwise called RDF vocabularies, intended to structure RDF resources. These resources can be saved in a triplestore to reach them with the query language SPARQL [5]. Each of the elements can be assigned with a weightage to denote the level of importance.

4. RELEVANCY FACTOR(S) IN KNOWLEDGE DISCOVERY AND ENRICHMENT

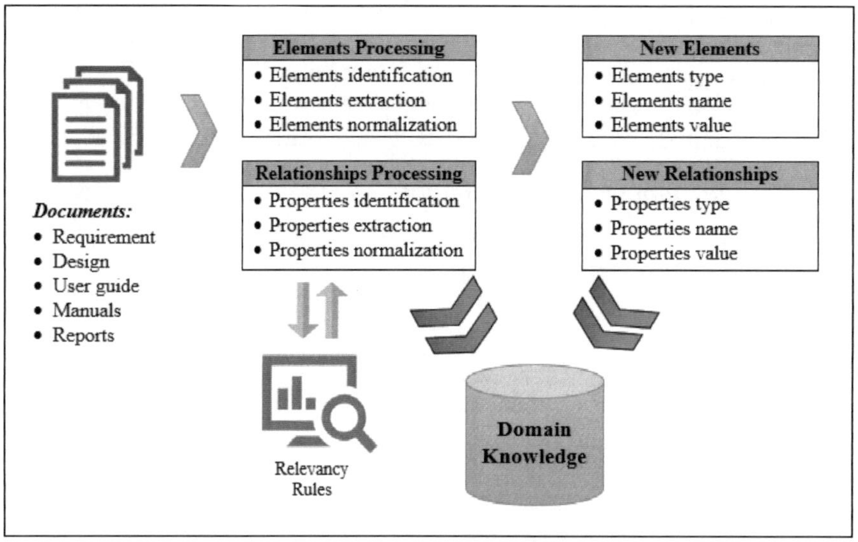

Figure 4. Elements and Relationships Processing with Relevancy Rules

Figure 4 above illustrated the elements and relationships processing with relevancy rules. Project documents will go through both elements and relationships processing to discover new knowledge [6]. Elements processing is done on the basis of parsing all the project documents to retrieve potentially relevant sentences to identify, extract and normalize the elements (subject and objects). Nevertheless, relationships processing will identify, extract and normalize the relationship (properties). Before adding the new elements and relationships into the domain knowledge, a set of relevancy rules will be applied on the retrieved information to eliminate irrelevant data and at the same time retain important knowledge.

4.1. Relevancy Rules

Relevancy factor(s) is important to determine the closely related elements area more relevant than distant elements. The relevancy rules consists of a list of relevancy factor(s) to extract both elements and properties relationships from documents. Each of the factor(s) is assigned with a weight. Ranking of the documents will be taking into consideration. The higher the ranking represent the level of importance of the documents and the elements extracted inside the documents will be higher in relevancy. As described below in Figure 5, the example of constructed top level of SDLC domain knowledge. It is an example listing of possible view point modeling in Design and Development stage. Not all of these views may be used for any one project and other views may be defined as necessary. Note that for some analyses elements from multiple viewpoints may be combined into a new view, possibly using a layered cross-referencing representation [7].

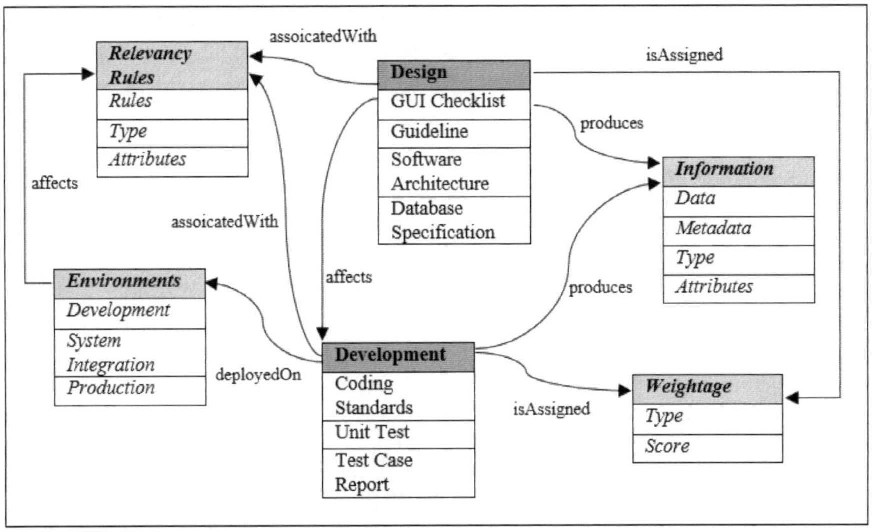

Figure 5. Example of Constructed Top Level of SDLC Domain Knowledge

4.2. Data Analytics in SDLC

The continuous processes generate data that meets the definition of big data. Most of the data gathered during the continuous processes hold a potential goldmine of information that can assist in decision making in each of SDLC stages. For example, an analysis of the data might help you to balance between practicing expensive and comprehensive testing to reduce the risk with better code coverage versus practicing a more efforts and cost-effective unit testing during development stage.

4.3. Benefits of Data Analytics

Data analytics technologies and techniques are widely used nowadays to ease organizations to do better decision making. It allows the organizations to reassess how they can work across the SDLC, and it can be applied to every stage of the software development as well as software delivery pipeline. Everyone involved in the process generates useful data. Whether it is the developer's bugs fixing report or tester's report as well as project meeting minutes, all the data will be useful to enable organizations to make better business decisions.

In future, smart systems may even work independently cross-referencing from historical decisions and suggesting optimizations based on historical data. The organizations can consider to incorporate machine learning on the historical data to discover hidden knowledge and relationships. Sooner or later, machine learning and big data analytics will be used across organizations to build and deliver software faster and better than ever before.

5. CONCLUSION AND FUTURE RECOMMENDATION

This paper elaborated the process of how useful knowledge (which also known as "elements" in this paper) can be extracted in each of the stage of software development life cycle with the essential of relevancy factor(s). The main advantages of knowledge discovery and enrichment are ease of reuse and expansion. However, areas of concern such as quality of the result from relevancy factor(s), attitudes of the users (Has this process caused organizational changes) and the frequency of the knowledge enrichment (cost involve of knowledge updating) play a significant roles. In order to build a comprehensive domain knowledge for this, exact and similarity of various software processes should be taking into consideration as the input sources.

ACKNOWLEDGMENTS

We would like to express our appreciation to all the people who have directly and indirectly guided me in completing this paper. We would also like to show our gratitude to our company MIMOS Berhad for the financial support. Last but not least, we sincerely thank the iCMLDE organizer for giving us this opportunity to share and publish our work internationally. Thanks for making this happen.

REFERENCES

[1] Alan Dennis, Barbara Haley Wixom, David Tegarden (2015). Textbook Systems Analysis and Design: An Object Oriented Approach with UML, 5th Edition, Accessed via http://as.wiley.com/WileyCDA/WileyTitle/productCd-EHEP003169.html

[2] Knowledge management systems life cycle, Accessed via http://www.cs.unibo.it/~gaspari/www/teaching/slides_KM2.pdf

[3] T. A. El-Diraby, C. Lima, B. Feis. (2005). Domain Taxonomy for Construction Concepts: Toward a Formal Ontology for Construction Knowledge. Journal: Computing in Civil Engineering, 4(394), 394-406.

[4] T. R. Gruber. (1993). A Translation Approach to Portable Ontologies. Journal: Knowledge Acquisition, 5(2), 199–220.

[5] Wikipedia, Accessed via https://en.wikipedia.org/wiki/RDF_Schema

[6] Ming Song, Wong Chul Kim, Dahee, Go Eun Heo, Keun Young Kang (2015). PKDE4J: Entity and relation extraction for public knowledge discovery, Journal of Biomedical Informatics, 320-332 pp, Accessed via http://www.sciencedirect.com/science/article/pii/S1532046415001756

[7] Wikipedia, Accessed via https://en.wikipedia.org/wiki/Ontology_engineering

Proc. 1st International Conference on
Machine Learning and Data Engineering (iCMLDE2017)
20-22 Nov 2017, Sydney, Australia

Query Optimization in Search Engines

Baisakhi Chakraborty[1], Nirjhar Datta Chaudhuri[2], Dambar Chetri[3] and Akshay Naik[4]

[1]Asst. Professor, Department of Computer Science, NIT, Durgapur, India
[2]Student, Department of Computer Science, NIT, Durgapur, India
[3]Student, Department of Computer Science, NIT, Durgapur, India
[4]Student, Department of Computer Science, NIT, Durgapur, India

Corresponding author's E-mail: baisakhichak@yahoo.co.in

Abstract

This paper aims to take a look at the way a search engine works when a user query is posted. The query processing part involves optimization of the user queries so that unnecessary and redundant parts of the query may be removed. Only the keywords are forwarded to the next stage of searching, which decreases the length of the posted query as well as the load on the further stages to increase throughput as well as response time. A number of linguistic aspects like methods of optimization, (namely stop words and name words that can be deleted without any change to the meaning of the original query) have been discussed in this paper. The spellchecking of the user query has been looked upon so that it is lingual and in sync with an English Dictionary.

Keywords: processor, optimization, stop word, name word, spell check.

1. INTRODUCTION

Information Retrieval (IR) works on finding the words or symbols through thousands of predefined strings of texts that match a user's query by Kraft H. Donald et al (2017). Any IR system aims to retrieve information based on the user's requirement. Query processing is a major task in IR systems that uses Corpus of information which is the backbone of any IR system by Goel K and Bhatia P (2016). The main activity of IR systems is query processing and retrieving information from a number of information resources by Lujia M et al (2017). IR systems as in Liddy E (2001) also known as search engines, are the focus of this project. The query optimization procedure adopted in this project aims at linguistically analyzing and modifying the query to improve the processing time and quality of the IR system, while saving on resources such as memory and process cycles. The IR systems can be broadly broken down into, a document processor, a query processor, searching and matching facility, and a ranking mechanism as in Murata T (2013). The IR systems works in conjunction with referential documents known as indexes, generated by the document processor. The generation of these indexes bear similarity with the process of optimizing and generating queries, by the query processor. The searching and matching facility connects the two processes; matching the results in the index against which the query was made. The ranking mechanism rounds up the process, adding the element of relevance, frequency, and other parameters that bear significance to the reliability of the system. The document processor normalizes the document, breaks it down, isolates and meta-tags it to form data structures that can be accessible to all the further processes. Furthermore "tokenizing" terms, deleting stop words and term stemming are all steps taken by the document processor towards the eventual extraction of index entries and building the inverted file which becomes the index. The details have been greatly discussed by Esbitan Samira SY (2012) and Stanford University Press (2008). The document processor also computes weights in order to make the entries more relevant and substantial.

2. RELATED WORK

Query Optimization in Search Engines is a widespread topic with efforts being made to improve timings and load decrease in processing the query. Advanced methods like making the system autonomous where it can learn to improve itself based on self-research are also being implemented.

In the research paper of Huston S and Bruce Croft W (2010), it has been mentioned how the removal of stop words, name words and stop structures can help to optimize query processing. It is further extended to the use of Classifiers CRF++ and YamCha for the purpose of sequential tagging of Stop Structures.
The main aspects of linguistic and grammatical concepts of Query Optimization are as follows:

1. Orthographic -- checking for typos, official variants (e.g., German/Dutch spelling), etc.
2. Morphologic -- including all forms of a given word via linguistic normalization (lemmatization).
3. Syntactic -- entity or phrase extraction, anti-phrasing, removing word-sense ambiguity (orange colour vs. fruit), etc.
4. Semantic -- applying a combination of general and specific thesauri and ontologies, automatic phrasing, etc., to understand the intention of the query.

3. ARCHITECTURE OF PROPOSED SYSTEM

Intuitive IR systems in FAST Search Best Practices (2006) allow for a fair amount of abstraction handling vague and misspelled queries, and even filtering and post processing results for a better navigation in results. The purpose of a search engine is to understand the context of the user's intent and to model the search around it, thereby returning the most relevant search results. The proposed work is using a rule based optimization technique. Each step is describing a single optimization rule.
A query processor follows seven distinct steps, some of which can be bypassed. They are tokenizing, parsing, and deletion of stop words, term stemming, query creation, query expansion and query weighing. As is apparent, query processing shares some steps with document processors as mentioned in Rosario B (2006) and Ioannidis Yannis E (1996).

a) Tokenizing is carried out on the query string in order to make it consistent and meaningful.
b) Parsing deals with determining the semantics that the queries intend to convey, using logical operators, Boolean operators and special symbols which would bear little to no relevance, otherwise .
c) Stop word removal is a focal point in the process of optimization.
d) Certain words share the basic structure among themselves e.g. analyzed, analysis, analyze share the prefix analy-. Grouping them under the same sub structure i.e. "analy-" helps broaden the search while also reducing the space need to store the individual terms
e) Query creation is a process of restructuring used to encompass all possible meaningful interpretations of the submitted query.
f) Query expansion incorporates synonymous interpretations of the query to improve the quality of the search.
g) Again query weighing is a practice which is used to obtain greater relevance in the search based upon parameters guarding the users' intent.

Query optimization by Slawski B (2009) is a process intended to make the context of the users' intent clearer to improve the quality of the search. Though supplementary, this phase of the query processing carries its worth through the resources that it helps save. Certain methods used in the optimization process are methods of suggestion. The method of query suggestions pulls up a list of candidate queries, and calculates their weight based on relevance and frequency. Upon completion the candidate queries are presented as a clustered list, based on rank. Also, the relevance of the suggested queries is specified. Suggested searches for related and frequently appearing queries are shown to searchers with the intention of making searching a better experience for the people who use search engines. The structure of suggestions that are presented can be decided upon various considerations such as popularity, subject matter, frequency, key words etc. The objects of considerations in this form of optimization are key words, their combination, popular phrases, and functionally or semantically similar suggestion cluster terms.

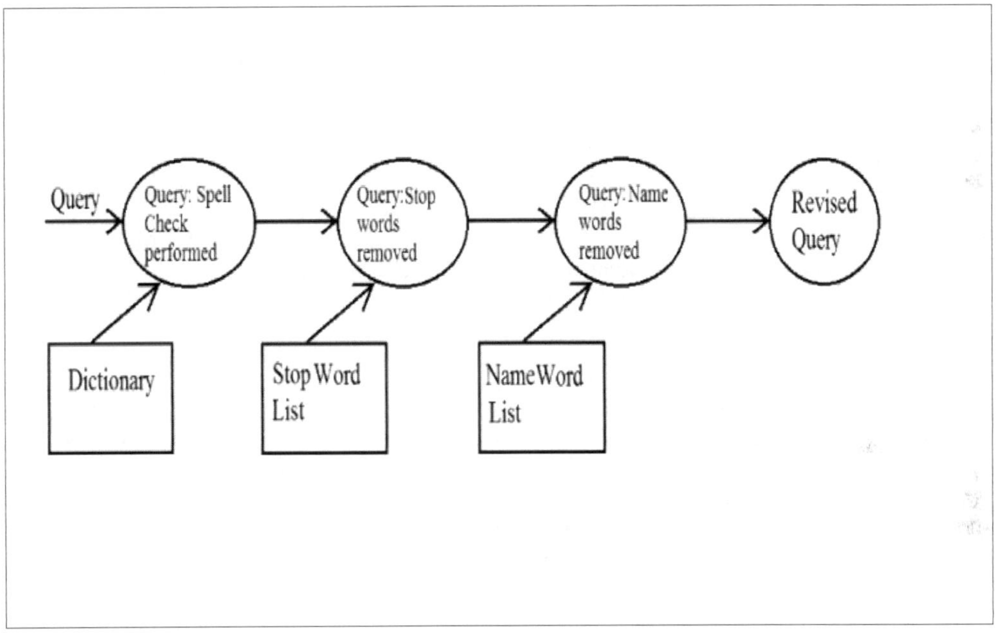

Figure1. Data Flow Diagram of Optimization Techniques used

3.1. Stop Words Removal

Stop words by Natural Language Processing department (2008) are extremely common words which do not serve any distinguishable purpose in the query processing function of an engine. The filtering out of such words plays an important role in query processing since it reduces the size of the query thereby reducing the size of storage and processing time needed for this purpose, without significantly affecting the quality of the results obtained. Search engines maintain a list of such words which they use as a reference in processing the query string. This list is not fixed or standard for a search engine and varies from engine to engine.
For example:

"Can I work while study in Europe?"

This technique produces the query:

"Work study Europe"

For removing the stop words from anywhere in the string, initially a separate text file is made with all the stop words. They are stored such that the actual source code does not need to be accessed every time a stop word has to be added or deleted from the original list. 25 words are chosen from a report as in Natural Language Processing department (2008).
After the stop words are stored, the next stage is to add them to the source code for matching and subsequent removal from the original string. Each of the values in the file are tested and then stored separately as words in an array. Then for each value we encounter in the aforementioned array, the original string is tested and if there is a match with any stop word in the list, the stop word is removed and replaced with a simple space.

3.2. Name Words Removal

Name words also add to the load of a query by adding unnecessary and unrelated question words at the start of a query. For example:

"Where is Delhi?"

This technique produces the query:

"*is Delhi?*"
And a combination of the above two techniques further shortens the query to

"Delhi"

One important factor that is a major difference between stop words and name words is that the name words are only present at the start of questions and hence the name words at the start of the query should be considered and not the ones in between as they may hold a separate meaning, different from a pure name word. The initial stage for Name word removal is the same as that for a Stop word removal, where a separate text file is made to avoid changes to the original source code. Then the name words present in the file are imported and extracted one by one onto an array. Then, they are tested with the original query and if there is a match, the position of the name word within the query is saved. If the position is zero, that is the name word is present at the initial position, then it satisfies the criteria of a name word and hence is removed with the help of the left trim php command ltrim.

3.3. Spellcheck

Spell check in Phpspellcheck (2004) at the GUI end is one of the foremost applications in query processing. Having a corrected query, unless it is a specific name, helps in finding the intended query correctly rather than having a redundant result as a consequence of incorrect spellings. For the purpose of checking spelling, a PHP code known popularly as "Phpspellcheck" has been embedded. It is a very useful code segment that considers a particular text area. The text within it co-relates it with a predefined English dictionary and provides alternatives to the incorrect word and even allows the user to choose among a number of possible alternatives if more than one appears. Phpspellcheck is an easy to use PHP Spell Checker script for websites and intranets. The Phpspellcheck component adds fast, reliable spell-checking to websites and intranets. Installed on either IIS or Apache Server, PHP Spell Check works on all major browsers since IE5. PHP spell check provides "As-You-Type" spellchecking with red-squiggly-underlines as well as a more traditional "Spell-checking Dialog" popup.

4. IMPLEMENTATION

In this paper, the Phpspellcheck in XAMPP-Apache Friends (2015) code has been applied to the GUI page rather than the processing side such that the user can check the spelling on words he has written incorrectly and correct them instantly. The "include.php" file is needed and is imported as this file contains the basic necessities for the implementation of the entire process. The contents of the text area entered in the form are considered for correction by indicating that in the specifications of the imported php file. The final implementation requires hosting the php files, along with the embedded spellchecker, so that they can be accessed and the subsequent files can be found in a single folder location. For this purpose, a XAMPP server has been used that hosts the entire project on port number 80 of the localhost, which is 127.0.0.1. XAMPP, along with WAMPP, is a very useful tool for this very purpose. XAMPP is a free and open source cross-platform web server solution stack package, consisting mainly of the Apache HTTP Server, MySQL database, and interpreters for scripts written in the PHP and Perl programming languages. The entire PHP based implementation of this project can be uploaded in a web server such as Apache HTTP server, WampServer, etc. This PHP based project has been implemented on Intel Corei3 based machine with 4 GB RAM and 500 GB Hard Disk.

5. RESULT

Since most of the applications of a search engine involve the user loads it initially and provides the query accordingly, hence we can say that the first case mentioned in the above table is the most frequent use, plus bringing in the entire page initially uses up comparatively more time than backtracking or refreshing.

Table 1. Time calculation of initial page load with embedded dictionary

Sl. No.	Page load situation	Time(in ms)
1	Initial loading of the page	316
2	First refreshing the page	60
3	Nth refresh of the page	18
4	Backtrack from 2nd page to 1st	15

Table 2. Query processing and result display

Sl. No.	Query situation	Time (in ms)
1	Small query (<10 words) without stop word and name word	1
2	Small query (<10 words) with stop word and name word	1
3	Medium query (100 words) with stop word and name word	1
4	Large query (1000 words) with stop word and name word	2

Thus, we have concurred that the time required for processing the query by subjecting it to the stop word and name word list is very negligible, hardly going past a couple of milliseconds at the most. Maybe with a wider stop word list the time will increase, but when compared to the time required for loading the dictionary in place, it will always be negligible to that.

6. CONCLUSION

Thus, the project we have conducted for the purpose of quantization in search engine query processing appears to be fruitful and the timings we have generated by using the timer and applying the various quantization techniques seem to be within limits and do not add much of an amount to the overall timing of the search process.
Nevertheless, the steps we have conducted are very preliminary steps in comparison to the vastness of all query optimization techniques. Yet our work is completely fault proof and in sync with the original objective of query optimization.

7. REFERENCES

Kraft H D, Colvin E, Marchionini G (2017). Fuzzy Information Retrieval, Synthesis Lectures on Information Concepts, Retrieval, and Services, 1, 63 pp.

Goel K, Bhatia P (2016). Information retrieval system using UNL for multilingual question answering. 2016 IEEE International Conference on Recent Trends in Electronics, Information & Communication Technology (RTEICT), 888 pp.

Lujia M, Wei B, Wugedele B, Wuriga Y, Tao H, Xiaobing Z (2017). A Mongolian Information Retrieval System Based on Solr. 2017 9th International Conference on Measuring Technology and Mechatronics Automation (ICMTMA), 335 pp.

Esbitan Samira SY (2012). A Personalized Context-Dependent Web Search Engine Using Word Net (Sama Search Engine), Islamic University of Gaza.

Huston S, Bruce Croft W (2010). Evaluating Verbose Query Processing Techniques, Centre for Intelligent Information Retrieval, Department of Computer Science, University of Massachusetts, Amherst.

Ioannidis Yannis E (1996). Query Optimizayion. ACM Computing Surveys *(CSUR)*, 28 (1), 121-123.

Liddy E (2001). How a Search Engine Works. Director of the Center for Natural Language Processing Professor, School of Information Studies, Syracuse University. Searcher, 9.

Murata T (2013). The Mechanism of the Search Engine. http://wiki.c2.com/?MechanismOfSearch Engine.

Rosario B (2006). Applied Natural Language Processing, University of California, Berkeley.

Slawski B (2009). How search engines may decide upon and optimize query suggestions. (DOI = www.seobythesea.com/?p=2409)

Search Query Processing (2006). Business white paper, FAST Search Best Practices.

Stanford University (2008). Stemming-and-lemmatization, Stanford University Press.

Natural Language Processing department (2008). Dropping common terms: stop words, Stanford University, USA.

Features of Phpspellcheck (2004). http://www.phpspellcheck.com/Features.

Phpspellcheck (2004). http://www.phpspellcheck.com/Main_Page.

XAMPP-Apache Friends (2015). http://www.apachefriends.org/en/xampp.html.

Proc. 1st International Conference on Machine Learning and Data Engineering (iCMLDE2017) 20-22 Nov 2017, Sydney, Australia

3D building reconstruction based on line detection algorithm

Xuyongzhe [1] and Minhaz Uddin Ahmed [2]

[2] Department of Computer Engineering, Inha University, Incheon, South Korea

[1] Zhejiang University City College, Hangzhou, China

Corresponding author's E-mail: yongzhexu@hotmail.com

Abstract

Automatic accurate detection of 3D Buildings using Google Street view of a city is challenging due to the presence of various horizontal and vertical occlusions such as trees, cars etc. Using different algorithms in such cases e.g. Laser Scanning (ALS), Unmanned Aerial Vehicle and a Depth camera (UAV RGB-D) map reconstruction, Light Detection And Ranging (LiDAR), is very costly. In order to obtain the correct coordinate value, developers need to match with wrong coordinate information, such as, Point Cloud Data (PCL), Digital Surface Model(DSM), and Digital Terrain Model (DTM) data. In this paper, we propose a 3D Building Reconstruction algorithm that uses 2D images. In our experiment, we use Hough line detection algorithm of OpenCV computer vision library to find outline information in a graph of 2D position. We transform a 2D position to a 3D space position coordinate. 2D image building only provides the frontal view of a 2D image which is not enough to find the coordinate in order to model a 3D building. To solve this problem multiple camera view information is required. Street view can add different view 2D image when detecting noise or occluded object in a scene. Firstly, we locate the noise and divide the noise palace to find the building edge for building construction. Secondly, we detect the line in a building and construct it to the 3D world. Our proposed new algorithm is very efficient to detect lines and 3D object reconstruction in street view. Moreover, it can also use internet map API to simulate a smart city.

Keywords: 3D Building, Pattern Recognition, Computer Vision, SFM,

1. INTRODUCTION

3D City models became popular to urban planners, real estate companies, telecom industries and the general population (for navigation) by Anastasios L. Kesidis, Nikolaos Vassilas, Theocharis Tsenoglou, Djamchid Ghazanfarpour (2015) and L Ragia, F Sarri, K Mania (2015). Popular search engines such as Google, Baidu, Naver, and other companies such as Apple, have researched into three-dimensional city map building, where 3D reconstruction information of buildings coordinates are created automatically without human labor involvement. In order to decrease the human resource association, computer vision technology was used to reduce the participation of individuals. This paper is based on the core content for street view image resources. Satellite picture matches with the GPS coordinates and identifies the three-dimensional reconstruction method of the real world of the building. Traditional building reconstruction method suchs as LiDAR (Light Detection And Ranging) by Baohua Chen, Lei Deng, Yueqi Duan, Siyuan Huang (2015), Mahdi Javanmardi, Yanlei Gu, Ehsan Javanmardi, Li-Ta Hsu, Shunsuke Kamijo (2015) or 3D scanning hardware is very costly due to the (point cloud Library) data, DSM data and DTM data by Lihua Tong, Manchun Li, Yanming Chen, Yafei Wang, Wen Zhang, Liang Cheng (2012) andMin Li, Leong Keong Kwoh, Chin-Jung Yang, Soo Chin Liew (2015). One of the major challenges is to organize this data to identify the coordinate values from data reorganization and modify the incorrect coordinates. Reorganization of incorrect

coordinates delays the development time and increases the human labor cost by Rakesh Kumar, M. G. Petovello (2015).
The reverse order of structure, which reduces cost, is a focus of this paper. In our proposed line detection method, we mostly used open source software platforms, which reduced the development cost. On the other hand baidu API made things effortless for the end user. Therefore, our proposed method is cost effective for development, suitable for enhancement and maintenance. The problem of PCL is the result of large-scale 3D coordinate point data. It is difficult to use the actual environment and modify the coordinates for the multi-point results, which increases the cost of modeling by Laura Micoli, Gabriele Guidi, Davide Angheleddu, Michele Russo (2013).

Figure 1. Shows the Baidu street view face watermark

2. RELATED WORK

Baidu Map service includes location sharing features and Street view API which allows users to swivel 360. The Baidu Map module also provides wrapper API which allows other modules or APIs to integrate with Baidu Map. The Baidu Geofield module provides Baidu Geocoder with a display textual address in China. The main difference with Google is Geocoder Chinese character support.

WGS84 : The world geodetic system is a standard coordinate system that was established in 1984. New WGS are available in the form of improved data and increased data coverage. An outstanding new satellite radar altimeter. Design least squares method called different types the Earth's gravity field. It is based on Global Positioning System (GPS).

GCJ02 : (aka Mars Coordinates) is a geodetic datum formulated by the Chinese State Bureau of Surveying and Mapping and based on WGS-84. It uses an obfuscation algorithm by Donoser, M, Schmalstieg, D. (2014). which adds apparently random offsets to both the latitude and longitude, with the alleged goal of improving national security.
A marker with GCJ-02 coordinates is displayed at the correct location on a GCJ-02 map. However, the offsets can result in a 100 - 700-meter error from the actual location if a WGS-84 marker (such as a GPS location) is placed on a GCJ-02 map or vice versa.

BD09 : BD-09 is a geographic coordinate system used by Baidu Maps, adding further encryption to the already encrypted GCJ-02 "to better protect user's privacy. Baidu provides an API call to convert from Google or GPS (WGS-84) coordinates into Baidu coordinates by Tabia, H, Laga, H, Picard D,Gosselin, P.H. (2014) Similar to GCJ-02, there are no APIs to convert in the other direction, but open source implementations in R Dimitrios Chrysostomou, Nikolaos Kyriakoulis, Antonios Gasteratos (2010) and various other languages exist.

3. METHODS

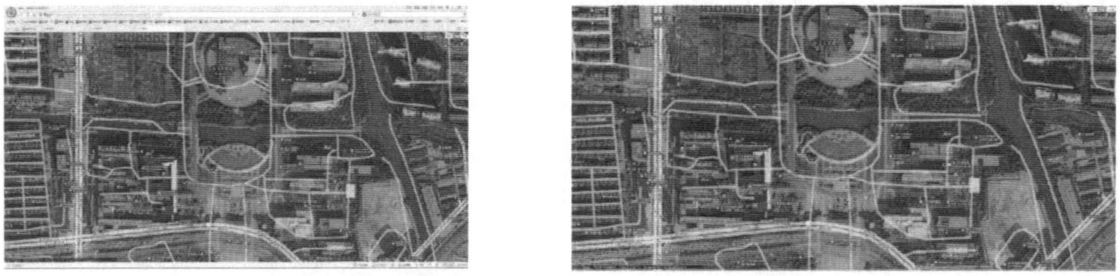

Figure 2. Seattle top view in baidu map API

Figure 2. shows a top view of Seattle City, China, using Baidu Map API. Baidu map API supports GPS address to find out buildings from street view image. We downloaded these map images of the same area in GPS localization path by Andrews, Rob W, Pollard, Andrew, Pearce, Joshua M (2013). First, we sett the top view image target building with GPS address, Web browser setting GPS value input into Baidu javascript map. Here centre and zoom (new BMap.Point(120.161502, 30.332646), 19); 120.161502 and 30.332646 is longitude and latitude 19 is zoom level. From Baidu Map API ak (primary access key) src="http://api.map.baidu.com/api?v=2.0&ak=primary access key similar to Google map API.

Figure 3. Top view of GPS addresses used canny edge and line detection

Second, find contour or line more than 3 points in the image to set the target building using OpenCV Canny edge detection algorithm. The few smaller lines are the GPS area outside.

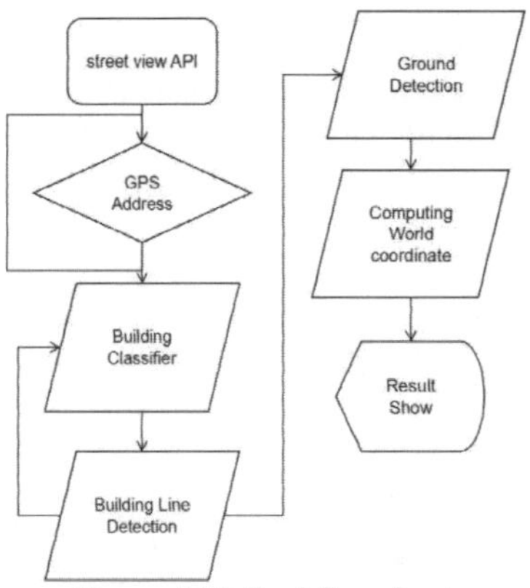

Figure 4.Method flowchart

Figure 4. shows the flow chart of our proposed algorithm. The GPS address was used to find street view images. The view angle requires the same target. From each location, a street view image was gathered in order to search the roof on the GPS coordinates classification into a building address.

The street view image of a building directly considered here with the coordinates to determine whether there is no classification to hide. The roof shape of a building and a street view of the segment shape structures are used together to compute the three-dimensional body estimation.

4. EXPERIMENT

Figure 5 and 6 show experimental results, downloaded, and captured from Baidu street view. These images were used to find line value. Pre-processes included bullring algorithms and produced complex image data which use corner detection in order to find Pilling of the line. These three lines make the 3D building corner.

Figure 5. Automatic capture of GPS street view and line detection

Figure 6 shows building reconstruction based on 20 street view images being overlapped. After line detection, and automatic 3D building is created. These lines have left right and longest value. And three lines have one corner point which is pilling of the line.

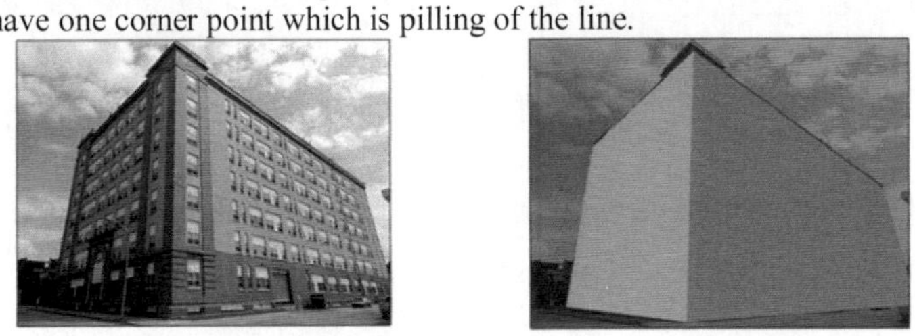

Figure 6. 3D Building Reconstruction

5. EXPERIMENTAL RESULT

In our experiment we used OpenCV 3.1 library with the integrated development environment visual studio 2015 and Microsoft .Net Framework 4.5. Our implementations were performed on two computers wither operating systems being Windows 7 and Windows 10, intel I5 and I7 CPU, primary memory 8Gb and 16Gb respectively. We used both programming languages python and C++.

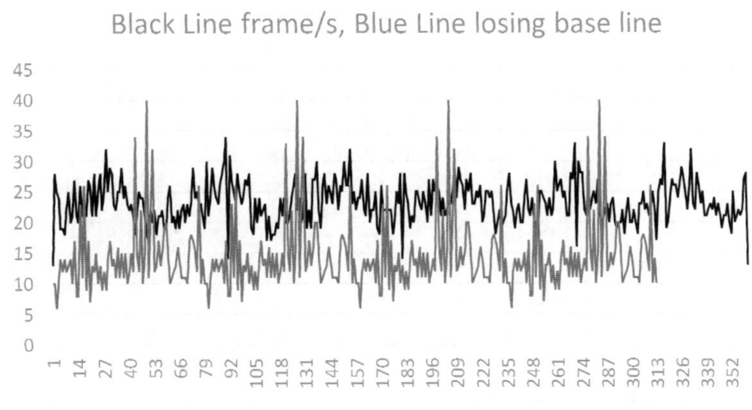

Figure 7. Frame performance and error rate of losing base line

Figure 7 shows frames with base line where the black line is the number of frame from Baidu MAP API and the blue line represents the error rate of losing base line. Frequency curves shows high and low line detection accuracy.

6. CONCLUSION

In this paper, we propose a cost effective way for 3D building reconstruction based on line detection algorithm. Our technique relied on street view map images and coordinate intersections. It is very time consuming and challenging to reconstruct three dimensional coordinates from two dimensional images.

7. ACKNOWLEDGMENTS

This work was supported by Zhejiang University City College Research Grant in 2017

8. REFERENCES

Anastasios L. Kesidis, Nikolaos Vassilas, Theocharis Tsenoglou, Djamchid Ghazanfarpour (2015). Experimental sensitivity investigation of mean-shift variants to bandwidth parameters for 3D buildings reconstruction, 9th International Symposium on Image and Signal Processing and Analysis (ISPA)

L Ragia,F Sarri,K Mania (2015). 3D Reconstruction and Visualization of Alternatives for Restoration of Historic Buildings - A New Approach,International Conference on Geographical Information Systems Theory

Baohua Chen, Lei Deng, Yueqi Duan, Siyuan Huang (2015). Building change detection based on 3D reconstruction, Jie Zhou Image Processing (ICIP), IEEE International Conference

Mahdi Javanmardi, Yanlei Gu, Ehsan Javanmardi, Li-Ta Hsu, Shunsuke Kamijo (2015). 3D building map reconstruction in dense urban areas by integrating airborne laser point cloud with 2D boundary map, IEEE International Conference on Vehicular Electronics and Safety (ICVES).

Lihua Tong, Manchun Li, Yanming Chen, Yafei Wang, Wen Zhang, Liang Cheng (2012).A research on 3D reconstruction of building rooftop models from LiDAR data and orthophoto, 20th International Conference on Geoinformatics.

Min Li, Leong Keong Kwoh, Chin-Jung Yang, Soo Chin Liew (2015). 3D building extraction with semi-global matching from stereo pair worldview-2 satellite imageries,2015 IEEE International Geoscience and Remote Sensing Symposium (IGARSS)

Rakesh Kumar, M. G. Petovello (2015). 3D building model-assisted multipath signal parameter estimation, Indoor Positioning and Indoor Navigation (IPIN), International Conference.

Laura Micoli, Gabriele Guidi, Davide Angheleddu, Michele Russo (2013).A multidisciplinary approach to 3D survey and reconstruction of historical buildings,Digital Heritage International Congress (DigitalHeritage)

Donoser, M, Schmalstieg, D. (2014).Discriminative Feature-to-Point Matching in Image-Based Localization, Computer Vision and Pattern Recognition (CVPR). Pp. 516-523

Tabia, H, Laga, H,Picard, D,Gosselin, P.H. (2014). Covariance Descriptors for 3D Shape Matching and Retrieval, Computer Vision and Pattern Recognition (CVPR). pp.4158-4192

Dimitrios Chrysostomou, Nikolaos Kyriakoulis, Antonios Gasteratos (2010).Multi-camera 3D Scene Reconstruction from Vanishing Points, Imaging Systems and Techniques (IST), International Conference on,2010, PP.343- 348

Andrews, Rob W, Pollard, Andrew, Pearce, Joshua M (2013). Photovoltaicsystem performance enhancement with non-tracking planar concentrators, Experimental results and BDRF based modelling, Photovoltaic SpecialistsConference (PVSC) IEEE 39th, pp.229-234.

Ken Sakurada, Takayuki Okatani, Koichiro Deguchi, (2013).DetectingChanges in 3D Structure of a Scene from Multi-view Images Captured bya Vehicle-mounted Camera", 2013 IEEE Conference on Computer Vision andPattern Recognition, pp.137-14

Dimitrios Chrysostomou, Nikolaos Kyriakoulis, Antonios Gasteratos (2010). Multi-camera 3D Scene Reconstruction from Vanishing Points, ImagingSystems and Techniques (IST), IEEE International Conference on,2010, PP.343 – 348

Tahmineh Partovi, Thomas Krauß, Hossein Arefi, Mohammad Omidalizarandi, Peter Reinartz(2014). model-driven 3d building reconstruction based on integeration of dsm and pectral information of satellite images, Geoscience & Remote Sensing Symposium , 2014 :3168-3171

Proc. 1st International Conference on Machine Learning and Data Engineering (iCMLDE2017)
20-22 Nov 2017, Sydney, Australia

A Survey on Cloud, Fog Computing and Internet of Things: An Architectural Viewpoint

Mohammad Sanaullah Chowdhury[1] and Ata Ullah[2]
[1]Department of Computer Science and Engineering, University of Chittagong, Chittagong, Bangladesh
[2]Dept. of Computer Science and Engineering, International Islamic University Chittagong, Bangladesh

Corresponding author's E-mail: sana1691@gmail.com

Abstract

It is the era of competitive technology, it is the era of cloud computing and smart world. The Internet of Things (IoT) provides the opportunity to connect billions of different types of every day's objects (things) with one another as well as with the Internet directly or through Local Area Network (LAN), paving a smooth and easy way to interact and share data and information. The primary concept of IoT is the pervasive presence of various types of wireless objects and embedded devices around us. However, the ultimate success of this IoT smart sensing environment depends heavily on a generally accepted, well-defined standard architecture that can provide a dynamic, scalable and secure framework for deployment of IoT. Full deployment of IoT and its applications in real life environment is not an easy task due to the limitations, challenges and security issues. IoT in collaboration with Fog computing and Cloud computing paradigm, revolutionizes the technology into a completely new dimension. In this paper, we discuss the employing and deploying of intelligent smart IoT devices in combination with Cloud and Fog computing paradigm. We also presented an improved and latency-minimized version of the Cloud and Fog centric IoT architecture. We then briefed on security and privacy issues along with challenges and future opportunities.

Keywords: Cloud Computing; Internet of Things; IoT; Fog Computing; Ambient Intelligence, Pervasive Computing.

1. INTRODUCTION

Nowadays, everything around us is going to be smart, can share data, information and services with one another. Of course this advancement facilitates improvement of public affairs. People can exchange information on the fly, can update themselves with the latest discount in shopping, instant traffic jam situation on the road, can access public health care info, can retrieve data and environment info from sensors, actuators and household appliances etc. All these are made possible due to the active participation of smart and intelligent devices comprising IoT. IoT is a dynamic network that connects smart physical objects (things) with high intelligence making use of the Internet. IoT devices doesn't necessarily need to be connected to the Internet, rather they can connect with each other through Bluetooth, ZigBee and NFC etc. You can control televisions, cameras, game consoles, cars, even washing machines from remote places. Your fridge can inform you about the kitchen-shopping list. The IoT goes far beyond that, with the help of sensor a farmer can know the conditions of the field by reading the data sent by the temperature and humidity sensor installed in the field [1].

Cloud computing, the next generational step in Information Technology [2], has emerged as the latest distributed computing model providing redundant, inexpensive and scalable resources on demand. As Cloud provides scalable storage and various computational resources IoT devices are able to implement a full scale ubiquitous computing environment [3]. On the other hand, Fog Computing is a

computing model similar to Cloud computing but at the edge of the networks and very close to the end users. It is specially created to focus on the IoT. As the IoT devices can transfer very huge amounts of data and computation offloading, and frequently need a very immediate response. Fog Computing paradigm is invented to work with IoT.
Rest of the paper is organized as follows. Section 2 reviews the background of IoT, Cloud and Fog computing systems. Section 3 explores the relevant architecture, while section 4 focuses on the security and privacy issues. Section 5 discussed the challenges and section 6 concludes the paper.

2. BACKGROUND

2.1. Internet of Things (IoT)

IoT is the concept of inter-connecting various devices and objects (things) together. The devices can be heterogeneous in nature and not necessarily needed to be connected to the Internet [4]. IoT devices are naturally connected via Wi-Fi, Bluetooth, ZigBee, Near Field Communications (NFC) and ZWave etc. The devices may be Smartphone, tablet, laptop, sensors, actuators etc. IoT devices can collect data and various sensitive information and share with other devices for further processes.

2.2. Cloud Computing

Cloud computing is a computing paradigm that provides scalable virtualized resources such as infrastructure, software, storage, database, servers and networks as per user's need via Internet technology. Based on the architecture and service providing system, there are 4 different types of service model, viz., 1) IaaS (Infrastructure as a Service), 2) PaaS (Platform as a Service) and 3) SaaS (Software as a Service). Cloud are becoming so popular and useful that private-public sectors are developing innovative technology to support industrialization, even in so many countries government facilities are moving towards the centralized cloud infrastructure.

2.3. Fog Computing

Fog computing is a paradigm similar to cloud computing typically located between the end users and traditional cloud data centers, but at the edge of the network [5] [6]. Fog computing also provides services such as data, computation, software and storage services on demand basis via Internet. Fog computing also enhanced Quality of Services (QoS) and location awareness for streaming and real-time applications. When IoT devices are connected to Fog server rather than Cloud data-centers, the latency and bandwidth can be reduced significantly as the Fog is situated at the vicinity [7] [8].

2.4. IoT with Cloud and Fog Computing

Due to the limitations of IoT devices, such as – battery life, computation, heat dissipation, limited storage – remote Cloud (Cloud computing) along with local Cloud (Fog computing) platforms are often employed to manage the data processing and storage needs. As the Cloud provides scalable storage and complicated computational resources big data processing, data analytics works, data mining and machine learning related resource-intensive processing can be performed seamlessly in Cloud-centric IoT. There are so many examples of implemented and concept-level projects using the IoT and Cloud and/or Fog as main communication platform. Some name of the projects and applications are:
- **Vehicular Fog Computing Networks** to deliver flyer ads and discount offers of the stores to the mobile users through WiFi.
- **Wearable devices** such as fitness trackers, smart watches, augmented reality glasses etc.
- **Wireless Sensor and Actuator Networks** (WSAN),

- **Smart Homes and Smart Buildings** projects which focuses on the citizens' issues by collecting application specific data, e.g. public parking monitoring; microclimate monitoring; access and mobility (pedestrian, cyclists, cars and freight vehicles), and real-time public transport data etc.
- **Cloud based Smart city** also focuses on Health Care Services (hospitals and personal care); City Safety and Environmental situations (noise, air and water quality); Sustainability and Utilities (better energy usage); Transport (personal and logistics); Smart Farming and many more [3].

3. ARCHITECTURE

IoT experts proposed various types of architecture such as, four-layer based smart object oriented architecture, five-layer architecture, SOA based architecture as well as middle-ware based architecture [9]. The three-layer architecture is shown in fig. 1(a), and five-layer architecture is shown in fig. 1(b). As IoT devices can be heterogeneous and distributed in large geographical area but are connected to each other comprising LAN through short range and low power-consuming services like NFC, Bluetooth etc., the Internet connectivity is done by use of any other access point such as Roadside Unit (RSU). To better grasp the working process and architecture of this emerging IoT field, one should understand its primary building blocks. Six primary building blocks of IoT are (1) Identification, (2) Sensing, (3) Communication, (4) Computation or Processing, (5) Services, (6) Semantics or Knowledge Extraction. We presented them in fig. 2.

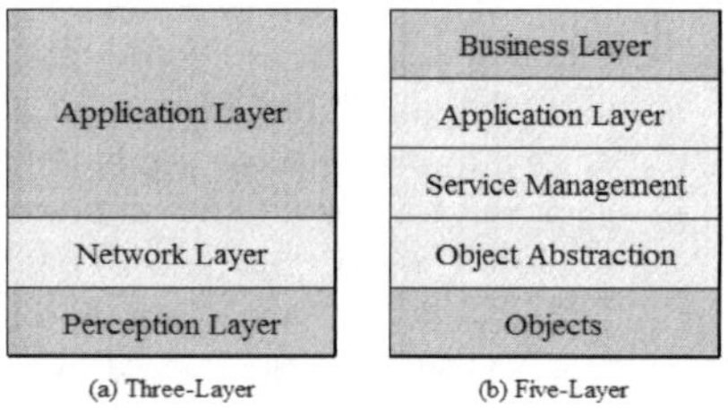

Figure 1. The architecture of IoT

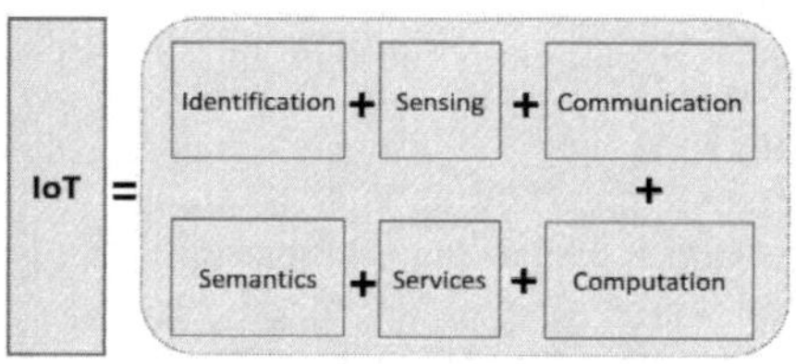

Figure 2. IoT elements

The end user device networks of IoT applications are dispersed in large-scale under geographically distributed network where devices are interconnected with Bluetooth, Zigbee, NFC, etc. Although these devices do not have direct Internet access, they keep generating application data and store it temporarily in local devices. Then they periodically send these data to the local cloud for further process. The existing architecture of IoT state that every things will send and receive data directly to and from Cloud data centers. But this is not possible for the local objects which has no Internet connectivity. As all IoT devices can communicate with each other via low-power-consuming technologies, placing a mobile device that can provide Internet facilities for communication with Cloud will help better for every type of IoT devices to function smoothly. [3] Introduced such a layer named as Mobile Access Points (MAP) for data collection and uploading process. MAP is a smart device with Internet access and travels among LANs of devices to collects all the application data generated by IoT devices and kept the data in a queue for uploading to the cloud.

As communication with the Cloud need stable Internet connection with higher bandwidth, moving IoT objects are often unable to achieve the expected result within due time. In some cases, such delay in communication process may be life threating (e.g. autonomous vehicles controlling system, vehicle-to-vehicle communication system or large-scale distributed rail network control system). That's why a closer to the end devices, low bandwidth consuming data processing system is needed. We deployed Fog platform for this purpose and proposed an enhanced architecture for low latency IoT environment with Fog computing. Fig. 3 depicted the roles of Fog computing and the cloud data-centers to deliver faster IoT services with low bandwidth. The proposed architecture is able to perform low latency computation and quickly store and retrieve data between end devices and Fog gateways. Fog gateways can play a vital role due to the following features [8]:
- Able to act as a mobile cloud as it is located at the edge of the networks and close to the IoT devices.
- Low setup price compared to the large cloud data-centers.

- Easily scalable and able to provide better performance for real-time interactive services.
- Fog servers can interoperate with all cloud providers with different deployment types.
- Smoothly perform data aggregation works and able to send partially processed result data to other Fog server or IoT objects instantly.

The deployment of Fog computing architecture between smart IoT devices and cloud platforms can increase the speed of computation services into manifold. Even big-data generated by billions of smart IoT devices can be transmitted and processed smoothly. We depicted the enhanced architecture in fig. 4, where all the end devices will comprise into various LAN according to their geographical locations. There will be a device such as Roadside Units (RSU) that will act as a Mobile Access Point (MAP). IoT devices will be able to connect and share the data with the MAP anytime via any low power wireless communication protocol such as Bluetooth, Zigbee etc. MAP will then perform offloading the data to the nearest Fog server for storage and computation.

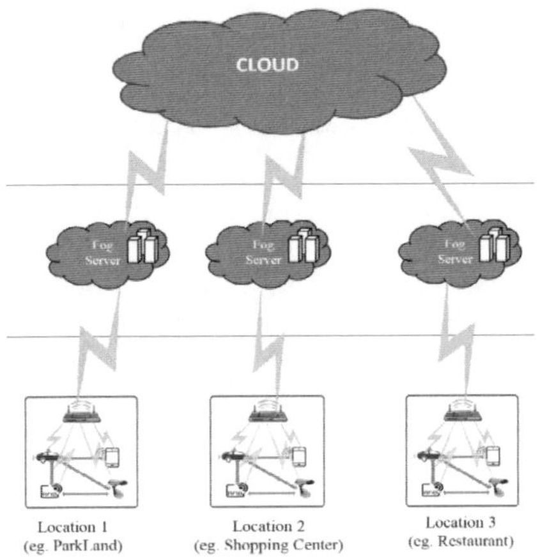

Figure 3. Roles of Fog Computing in IoT

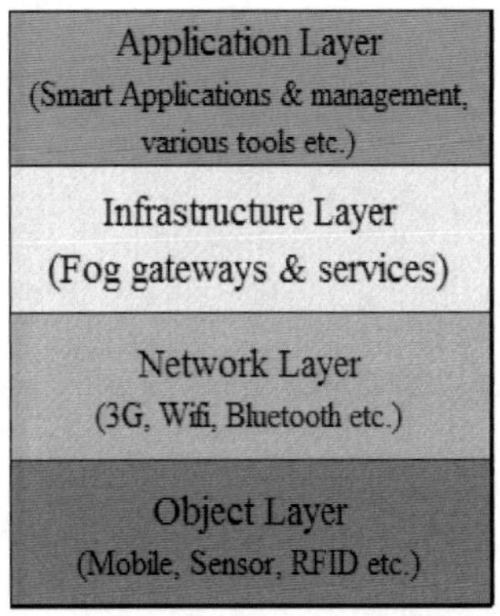

Figure 4. Four-Layer IoT architecture

4. SECURITY AND PRIVACY ISSUES

IoT is comprised of not only mobile devices and wearable objects, rather hundreds of thousands of sensors, integrated or embedded devices connected for a wide variety of tasks. The rapid growth of IoT faces different challenges and risk not only for the large companies and private sector, but also for the security and privacy of individual users of these devices. Visibility awareness and safety risk, public cloud vulnerability, security of confidential information, application security as well as authentication and authorization are most vulnerable fields targetable by the intruders. A recent study by HP [10] shows that 90% of devices collected at least one piece of personal information via the device, Cloud or its mobile application. 80% of devices raised privacy concerns while 70% of the devices does not encrypt the data, leading to increased problems of security. To make the IoT more secure, [11] recommended six actions for the companies, viz.,

i) Adopt a comprehensive framework, proactive, enterprise-wide strategy for securing the IoT.
ii) To run a full security audit to assess the complete IoT deployment (i.e. IoT devices, network infrastructure, mobile, web and cloud touch points etc.).
iii) Security procedures need to be instilled before deployment of the device.
iv) Mobilize the larger workforce (i.e. employees of product design, the supply chain, production and other traditional parts of the organization) around IoT security.
v) It need to ensure that customers, suppliers and others partners strictly adhere to standards of security.
vi) To change the role of IT as a skilled partner rather than service provider.

5. CHALLENGES

Efforts have been made by World Wide Web Consortium (W3C), Internet Engineering Task Force (IETF), EPCglobal, Institute of Electrical and Electronics Engineers (IEEE) and the European Telecommunications Standards Institute (ETSI) to create their individual protocol model for IoT, but unfortunately till now there is no common and generally accepted model protocol. Even though there are so many things to be considered as important for establishment process of IoT frameworks, making a standard IoT architecture and standardization of IoT protocol is very urgent. In this paper, we already explored and focused all the latest Cloud-oriented IoT innovative technologies, currently undergoing projects and concepts. Wide and fast spread of IoT applications in everywhere emerges several types of challenges both in technical and social arena. To adopt the future of IoT seamless and reliable, all these challenges needed to be overcome. The challenges currently facing by IoT are discussed below.

Availability: IoT functionalities must be available both in software and Hardware level. IoT applications must have the ability of providing services for everyone simultaneously at different places (i.e. everyone, everywhere and everything style).

Mobility: As IoT comprise with mobile devices, mobility challenge is highly concerned. For a moving object (thing), changing of fog gateway while processing services or transferring data might cause serious interruption or even loss of data.

The big data problem: Instant computation and real-time processing of the big data generated or collected by IoT devices is still a challenge.

Performance: Evaluation of heterogeneous IoT systems is still a very crucial thing as it depends not only on single system rather all the end devices, protocols and many other components and technologies.

Management: Managing billions of heterogeneous devices are an incredible task.

Reliability: For communication of IoT devices reliability is a very must important scheme without which processing of computation, transmission of data, and gathering of information may eventually lead to wrong direction forcing the ultimate result into unexpected or garbage.

Some more additional issues end users may face include but not limited to are: - privacy and trustworthiness issues, security problem, universal standard connectivity issue, lack of common architecture and lightweight protocol, power management etc.

6. CONCLUSION

The IoT is virtually making our environment ubiquitous systems integrating with Cloud and Fog computing paradigm. IoT have radically transformed many aspects of our daily lives. From a positive point of view, we have seen a wide range of applications and – from Nano-to-Mega devices, from shirt collar to bed, from home refrigerator to fire alarm – talking with every surrounding thing; making them smart and responsive. On the other hand, it imposes a great potential vulnerabilities and insecurities on privacy of individuals. Even though, several world class technology giants are working hard to make the IoT a well adopted secured form, it still need a lot of things to be done in order to fulfill the original vision of IoT. Lack of generally accepted model architecture is also hindering in progress of IoT's advancement. A common and standard model architecture, backward compatible system, and secure and reliable frameworks are needed to achieve the expected reality of IoT concept. In this paper, we have explored the existing architectures of IoT along with the IoT elements and building blocks generally deemed as well-adapted forms to suit IoT's requirements, and also provided an overview on proposed low latency architecture.

In future, we would like to implement a Fog computing oriented framework that will be able to interact with billions of smart devices from different manufacturers, of different throughput, having multiple communication protocols that can handle the various types of challenges.

7. REFERENCES

Tsai CW, Lai CF, Athanasios V. Vasilakos (2014). Future Internet of Things: open issues and challenges, Springer Science+Business Media New York.

Armbrust M, Fox A, Griffith R, Joseph AD, Katz R, Konwinski A, et al. (2010). A View of Cloud Computing, Communication of the ACM, vol. 53, pp. 50-58.

Yuan D, Jin J, Grundy J, Yang Y (2015). A Framework for Convergence of Cloud Services and Internet of Things, 19th International Conference on Computer Supported Cooperative Work in Design (CSCWD), pp. 349-354.

Carrillo E, Benitez V, Mendoza C, Pacheco J (2015). IoT framework for Smart Buildings with Cloud Computing, IEEE Smart Cities Conference (ISC2).

Al-Fuqaha A, Guizani M, Mohammadi M, Aledhari M, Ayyash M (2015). Internet of Things: A Survey on Enabling Technologies, Protocols and Applications, in IEEE Communications Surveys & Tutorials.

Stojmenovic I, Wen S (2014). The Fog Computing Paradigm: Scenarios and Security Issues, IEEE, Federated Conference on Computer Science and Information Systems, pp. 1–8.

Bonomi F, Milito R, Zhu J, Addepalli S (2012). Fog computing and its role in the internet of things, First Edition of the MCC Workshop on Mobile Cloud Computing, ser. MCC'12. ACM, pp. 13–16.

Yi S, Li C, Li Q (2015). A survey of Fog computing: Concepts, Applications and issues, ACM.

Babu SM, Lakshmi AJ, Rao BT (2015). A Study on Cloud based Internet of Things: CloudIoT. Global Conference on Communication Technologies (GCCT 2015), pp. 60-65.

Hewlett Packard Enterprise Security Research Report (2015). Accessed via http://www8.hp.com/h20195/V2/GetPDF.aspx/4AA5-4759ENW.pdf on 09-06-2016.

Securing Internet of Things: A report from The Economist Intelligence Unit & Hewlett Packard Enterprise (2016). Accessed via http://hpe-enterpriseforward.com/eiu-securing-iot/ on 08-07-2016.

Proc. 1st International Conference on Machine Learning and Data Engineering (iCMLDE2017)
20-22 Nov 2017, Sydney, Australia

Automatic Ear Detection using Deep Learning

MD Moniruzzaman[1] and Syed Mohammed Shamsul Islam[2]

[1]Postgraduate Student, School of Science, Edith Cowan University, Joondalup, WA 6027, Australia
[2]Lecturer, School of Science, Edith Cowan University, Joondalup, WA 6027, Australia

Corresponding author's E-mail: syed.islam@ecu.edu.au

Abstract

Ear based biometric identification can be the solution for instance such as surveillance where other biometric traits are simply very hard to access. Although many semi-automatic approaches have been made to detect ear and use it for human recognition purpose, most of them are based on feature extraction or shallow machine learning approaches. Very few approaches those used deep neural network architectures are either having less hidden layers, or a combination of deep neural network and feature extraction classifiers or already trained complex deep convolutional network. In this research approach, a deep but simple raw convolutional neural network have been used to detect ear from an ear and non-ear environment which is the initial part of ear based biometric implication. Data-Augmentation have also used and a comparative analysis have been done for original data set and Augmented dataset. Using this deep but from scratch architecture trained by 792 original images we achieved promising output which show larger data could achieve higher accuracy.

Keywords: *Ear detection, Deep Learning, Convolutional Network, Data augmentation.*

1. INTRODUCTION

During the course of last decade, researchers are focusing actively on biometrics research due to the growing need for automatic authentication of human individuals. In biometric identification, ear based authentication and human recognition proved itself a novel research field as suggested by Zhang and Mu (2017). The potential, credentials and possibilities of using human ear for recognition and human identification was proposed by Bertillon (1890). According to Jain et al (2007) and Omara et al (2017), to satisfy the traditional personal authentication properties, a biometric trait should be unique, universal, permanent and easily collectible. The uniqueness of the helix part of human ear has been proven by Iannarelli (1989). Not only helix, anti-helix, which is a parallel part of helix with a distinctive hairpin-bend shape also proved prominent by Hurley et al (2007). Human facial expression doesn't have any effect on ear, with age the change is negligible and the background is always predictable as ear is fixed firmly on the side of human head. Collection of ear image does not create a sense of anxiety and no hygiene issue raises as the collection process is touch-less. Moreover, as the size of the ear is significantly larger than other biometric traits like finger print, iris, retina, it proves more accessible while the incident is associated with criminal act. Present Artificial Neural Network based ear detection and recognition approaches can be broadly categorized in to 3D point cloud based approaches or 2D image data approaches. This research work deals with only 2D images.

One of the earliest approaches of using Neural Network based Ear recognition was experimented by Galdámez et al (2014). In their approach they used Speeded Up Robust Features (SURF) and Fisher Linear Discriminant Analysis (LDA) as the input of their two neural networks. On their later approach Galdámez et al. (2016) used deep Convolutional neural networks (CNN) based ear recognition system to identify a person by the ear image. Tian and Mu (2016) designed a neural network having three convolutional layer followed by a fully connected layer and a soft-max classifier. They used University of Science and Technology, Beijing (USTB-III) dataset. Emeršič et al. (2017) tried to train convolutional neural networks with small training data. They implemented aggressive data augmentation with selective model learning technique for limited training data. For the solution of lack of training data problem for deep learning based technique Omara et al. (2017) proposed a technique called pairwise SVM. Their neural networked was based on VGG-M Net. Along with pairwise SVM, they also used Principal Component Analysis (PCA) so that the dimension can be reduced before classification. They used USTB-I and USTB-II dataset for their experiment and Accuracy was more than 98%. Cintas et al (2017) proposed a new approach based on deep learning and Geometric Morphometrics for feature extraction and ear detection in the form of landmarks. They trained three different convolutional neural networks with a set of manually landmarked examples. They used a dataset (CANDELA, Consortium for the Analysis of the Diversity and Evolution of Latin Americans) of 2735 manually landmarked ear images. Zhang and Mu (2017) in their recent approach, proposed an efficient method based on Multiple Scale Faster Region-Based Convolutional Networks (Faster R-CNN) for ear detection from 2D images. For this experiment they created an ear dataset consisting 100 images collected from internet and are captured from real world situation and named it WebEar dataset. Along with this dataset they also experimented on UBEAR and UNDJ-2 (University of Notre Dame J2) dataset. All the above approaches are summarised into Table 1.

Table 1. Summary of Deep Neural Network based ear detection and recognition approaches

Author	Technique Used	Number of Image	Accuracy
Galdámez et al (2014)	SURF and LDA as input of Neural Network	300	97% for SURF and 93% for LDA
Galdámez et al (2016)	CNN Architecture with sliding window optimisation	52800	99.02%
Tian and Mu (2016)	3 Layer CNN	79 Subjects	98.27%
Emeršič et al (2017)	Aggressive data augmentation and Selective model learning	2304	>95%
Omara et al. (2017)	Pairwise SVM & PCA	488	>98%
Cintas et al (2017)	Deep Learning with Geometric Morphometric	2735	
Zhang and Mu (2017)	Faster R-CNN	10995	>98%

The very first step of ear based recognition system is automatic ear detection. The main intention of the project was to experiment and judge the possibility of using a raw deep convolutional neural network architecture based algorithm to detect ear in the small 2D data sets available for ear and maximising efficiency and reducing time to ensure practical use of the system. The outcome of the project will set the base for a fully automated, efficient and real time ear based human authentication system in the future work.

The organisation of the rest of the paper is as follows. The elaboration of the approached research design and methodology is discussed in section 2. Experimental results and discussion is provided at section 3. Finally conclusion is drawn and indication of future research is provided in section 4.

2. RESEARCH DESIGN AND METHODOLOGY

Three separate attempts have been made to detect ear from an ear and no-ear environment in this research approach using Deep Convolutional Neural Network. In this Section the architecture of the Deep Neural Network have been have been discussed along with the description of the dataset used and the Data Augmentation technique applied as well as training and testing methods.

2.1. Dataset

To create a perfect ear and non-ear environment for ear detection purpose, total five classes of images were selected for experiment one and dataset-1 was created. The five image categories involved digital photos of cat, dog, ear, horse and human. Cat, dog, ear and human every category have 202 images collected from github (https://github.com/anujshah1003/own_data_cnn_implementation_keras) and for ear 182 images were used from University of Science and Technology Beijing- I (USTB-I) [Omara et al (2017)]. So the whole dataset includes 990 images. For Second attempt the image class have been reduced from five to three that includes cat, ear and horse. With the previous 202 images for cat and horse, another 300 images were added from Canadian Institute for Advanced Research (CIFAR-10) database for each of the category to create dataset -2. And for ear 309 images were added from USTB-II and 11 photos were added from USTB-III. So for second approach 1506 images were used from 3 classes, each having 502 images. For third attempt, the size of the data set was 10574, including 3542 ear images, 3514 cat images and 3518 horse images. This dataset-3 is made up with all the images from dataset-2 and further data augmentation.

2.2. Data Augmentation

It is commonly accepted that the more data can be used during the Deep Neural Network training phase, the more effective it becomes for validation phase [Wang and Perez (2017)]. As for first and second attempts, the size of the data were fairly small, Data augmentation technique was used to increase the size of the total data more than ten thousands. The augmentation procedures that were performed in this experiment is listed below.

1. Rotation of up to 40° in both direction
2. Width sift from 0 to 20% range
3. Height shift from 0 to 20% range
4. Shearing from 0 to 20% range
5. Image Zoom up to 20%
6. Horizontal flipping

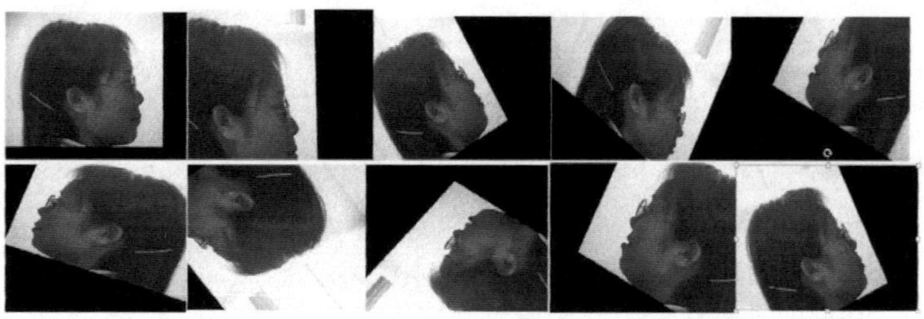

Figure 1. Sample Ear images after Augmentation

2.3. Network Architecture

To keep the consistency in the research, for all the attempts same architecture have been used. The architecture of the Deep Neural Network used in this project (https://github.com/anujshah1003/own_data_cnn_implementation_keras) is shown in the Table-3 below.

Table 3. Layers of Deep Neural Network Architecture Used

Layer Index	Type	Filter Size	Output Size
L1	Convolution2D	3 × 3	32 × 126 × 126
L2	ReLU		32 × 126 × 126
L3	Convolution2D	3 × 3	32 × 126 × 126
L4	ReLU		32 × 126 × 126
L5	MaxPooling2D	2 × 2	32 × 63 × 63
L6	Dropout		32 × 63 × 63
L7	Convolution2D	3 × 3	64 × 61 × 61
L8	ReLU		64 × 61 × 61
L9	MaxPooling2D	2 × 2	64 × 30 × 30
L10	Dropout		64 × 30 × 30
L11	Flatten		57600 × 1
L12	Dense		64 × 1
L13	ReLU		64 × 1
L14	Dropout		64 × 1
L15	Dense		5 × 1
L16	Softmax		5 × 1

2.4. Training and Testing

For all the three instances Datasets were divided into training set and test set. For dataset-1 training set contains 792 images and 198 test images. Dataset-2 was divided into 1204 training images and 302 test images, and dataset-3 was divided into 8459 training and 2115 test set. For all three attempts, input images were converted to 128 × 128 grey scale images. Training was done using Rectified Linear Unit (ReLU) and Softmax Activation function. Training batch size was set 16 as the training was done on a GPU less computation system. Number of epoch was 20 for all the attempts to keep the validation result trustworthy. The whole experiment was done using Theano backend Keras library on a python 2.7 based programming language. Open CV was used for image processing. The experimental setup used a computation machine having 2 GHz Intel Core i5 processor, 8GB RAM and an Intel 540 Iris Graphics card. So the whole system is really cost effective and Table 4 indicates the very less execution time.

3. RESULT AND DISCUSSION

From Table-4 we can see that the training accuracy for dataset one 86.11% but the validation accuracy is 56.56% which is fairly low. The reason behind the low validation accuracy is the very small size of the data. The size of the training data was only 792. When the data was increased to 1204 for training at dataset 2 and the number of class was decreased to three from five, the validation accuracy increased to 81.79%. So it is fairly obvious that the increase of main data increased the accuracy for the architecture.

Table 4. Training and Validation accuracy with loss for all the 3 datasets

Data Set	Execution Time Per Epoch	Training Accuracy	Validation Accuracy	Training Loss	Validation Loss
Dataset- 1	153 sec	86.11 %	56.56%	0.3920	1.3328
Dataset-2	235 sec	90.12%	81.79%	0.2616	0.3933
Dataset-3	1621 sec	87.21%	77.45%	0.3668	0.5812

Dataset-3 includes the augmented images of dataset-2. The training size of this third attempt was 8459 which is much larger than that of dataset-1 and 2. But the validation accuracy is 77.45% which is slightly lower than that of attempt-2. This results indicated that the augmentation procedure (shearing, flipping, zooming, rotation etc.) chosen for this experiment actually did not performed up to expectation. Comparison for the training and validation accuracy is presented in Figure-2.

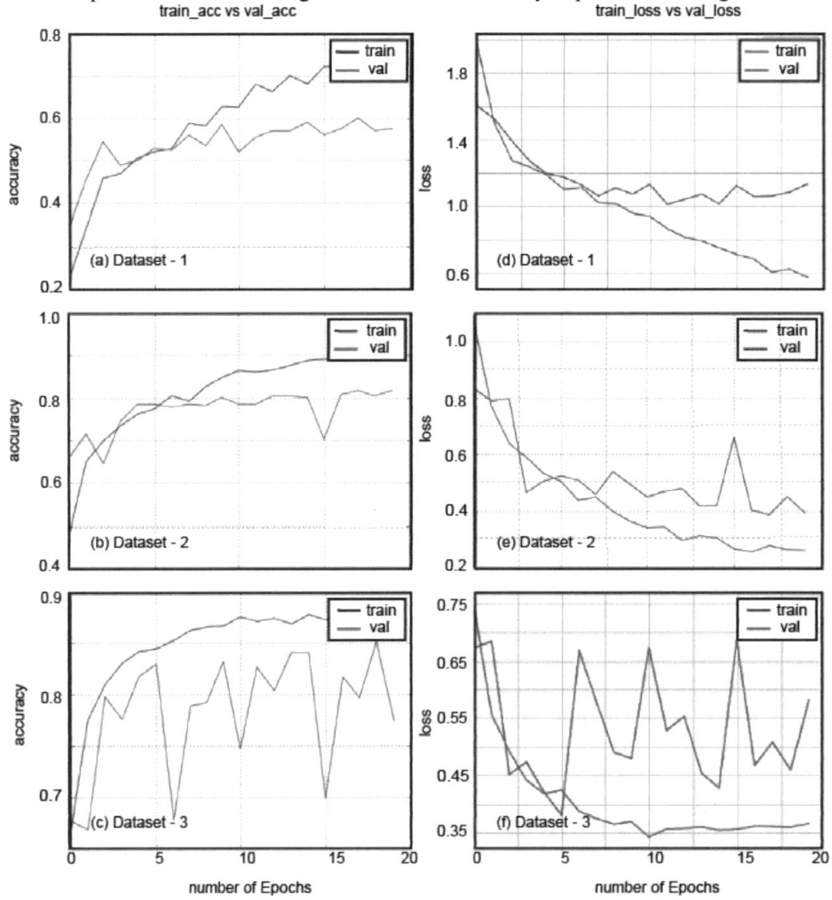

Figure 2. Train and Validation Accuracy with loss for dataset-1, 2 and 3 consecutively

For Neural Network loss is measured as negative log-likelihood which is not a percentage. The training loss and validation loss for all the attempts are presented in table-4 and Figure-2, which also indicate the correctness of the accuracy data. For attempt -2 for dataset-2 the network performed better than other two attempts.

4. CONCLUSION AND FUTURE WORK

In this research project deep learning architecture based ear detection was attempted in an ear and non-ear environment. The proposed architecture is capable of detecting ear from a mixed environment and the highest validation accuracy of 81.79 % achieved for attempt two with dataset-2 which indicates the success of the research project. The result of the experiment is considerable as the time period of the research project was very short (only 12 weeks) and experiment with deep neural network architecture takes a significant amount of work and time especially the network selection, data augmentation and training phase. The accuracy of the network would be much higher simply if the size of the original image data were larger. So the same experiment will be done after collection of more digital ear images from different data sources. This project tested the accuracy of a single deep neural network. Several other networks will be used in the future work so that a comparative analysis can be done and a more robust architecture can be selected. Regularisation, Network initialisation, batch normalisation and some other type of data augmentation technique will be used in future work which may help to increase the accuracy of the ear detection. Moreover, this experiment deals with the detection of ear images from a mixed image dataset, but the future research will be focused on the biometric implication of ears i.e., to detect a person from ear image which will address a real life problem.

ACKNOWLEDGMENTS

Authors would like to acknowledge that this work is done with the support of Australian Government Research Training Program Scholarship and Edith Cowan University ECR grant. All the ear data were collected from University of Technology, Beijing (USTB I, II, III). Non-ear data were collected from Canadian Institute for Advanced Research (CIFAR-10) and data available in Github.

REFERENCES

Bertillon A (1890). La photographie judiciaire, avec un appendice sur la classification et l'identification anthropometriques, Gauthier-Villars, Paris.

Cintas C et al (2017). Automatic ear detection and feature extraction using Geometric Morphometrics and convolutional neural networks, IET Biometrics, 6(3), 211-223.

Deepak R, Nayak AV, Manikantan K (2016). Ear detection using active contour model, IEEE Emerging Trends in Engineering, Technology and Science, 1-7.

Emeršič Ž, Štepec D, Štruc V, Peer P (2017). Training Convolutional Neural Networks with Limited Training Data for Ear Recognition in the Wild, IEEE International Conference on Automatic Face & Gesture Recognition, Washington, DC, 987-994.

Galdámez PL, Arrieta MAG, Ramón MR (2014). Ear Recognition with Neural Networks Based on Fisher and Surf Algorithms, International Conference on Hybrid Artificial Intelligence Systems, Springer, 254-265.

Galdámez PL, Raveane W, Arrieta AG (2016). A brief review of the ear recognition process using deep neural networks, Journal of Applied Logic, 62-70.

Ganesh MR, Krishna R, Manikantan K, Ramachandran S (2014). Entropy based Binary Particle Swarm Optimization and classification for ear detection, Engineering Applications of Artificial Intelligence, 27, 115-128.

Hurley DJ, Arbab-Zavar, B, Nixon, MS (2007). The ear as a biometric, 15th European Signal Processing Conference (EUSIPCO), Poznan, Poland, 25-29.

Iannarelli AV (1989). Ear identification, Paramount Publishing Company.

Jain A, Flynn P, Ross AA (2007). Handbook of biometrics, Springer Science & Business Media, 131-150 pp.

Omara I, Wu X, Zhang H, Du Y, Zuo W (2017). Learning pairwise SVM on deep features for ear recognition, IEEE International Conference on Computer and Information Science, Wuhan, 341-346.

Tian L, Mu Z (2016). Ear recognition based on deep convolutional network, International Congress on Image and Signal Processing, BioMedical Engineering and Informatics, Datong, 437-441.

Wang J, Perez L (2017). The Effectiveness of Data Augmentation in Image Classification using Deep Learning, Stanford University.
Yan P, Bowyer K (2005). Empirical evaluation of advanced ear biometrics, IEEE Computer Vision and Pattern Recognition-Workshops, 41.
Yan P, Bowyer KW (2007). Biometric recognition using 3D ear shape, IEEE Transactions on pattern analysis and machine intelligence, 29(8), 1297-1308.
Zhang Y, Mu Z (2017). Ear Detection under Uncontrolled Conditions with Multiple Scale Faster Region-Based Convolutional Neural Networks, Symmetry, 9(4), 53.

Proc. 1st International Conference on Machine Learning and Data Engineering (iCMLDE2017)
20-22 Nov 2017, Sydney, Australia

Automatic Damage Detection of Containers Based on Hierarchical Image Analysis

Youngjae Lim[1], Jang-Hee Yoo[1], and Kiyoung Moon[2]

[1]SW·Contents Research Laboratory, ETRI, Daejeon, South Korea
[2]Regional Industry IT Convergence Research Section, ETRI, Daegoo, South Korea
Corresponding author's E-mail: yjlim@etri.re.kr

Abstract

In this paper, we present a new method for analysing the damage to a logistics container by automatic analysis of its integrity based on parts-zone segmentation and hierarchical image analysis. To achieve this, we use a smart phone or tablet in order to acquire an image from the rear side of the container. The automatic damage detection system consists of three stages: i) dividing the image into small areas as minimum analysis units, ii) calculating the score of the individual units, and iii) summation of the scores for each unit to estimate the degree of integrity. In experiments, we have successfully evaluated the performance of the proposed method with data from a smart phone camera. As such, the proposed method provides automated damage detection which has the advantage of offering a relatively inexpensive and expeditious estimate of container damage.

Keywords: Logistics container, Damage analysis, Zone segmentation

1. INTRODUCTION

In general, the integrity of logistics containers is visually inspected by human experts at gates of container terminals or bonded warehouses. These visual inspections are carried out under adverse conditions such as vehicle fumes or bad weather, and are also based on subjective judgements that depend on the proficiency of the inspection personnel. Therefore, the need for an automated system capable of quantitative analysis is increasing. Recently, techniques for automatically checking damage to containers using 3D laser scanning devices such as the LaseCDI-GATE system (2014), which require fixed facilities including expensive equipment, have been applied. Huang et al (2013) have proposed the system for detecting damages by comparing two images of the containers acquired before and after the transfer using fixed cameras. However, this method only detects changes due to breakage during transport, and cameras of the same specifications should be installed at the departure and arrival points.

In this paper, we propose a method to measure the integrity of a container using only a mobile camera without additional expensive equipment. For this purpose, the rear side image of the container is captured by the camera on a mobile device such as a smart phone or tablet at the container inspection site, it is divided into parts-zones as the analysis units, and quantitative integrity parameters such as a contamination score, decrepitude score, and locking score in each parts-zone are calculated. Finally we can obtain the degree of integrity for the entire container from these integrity parameters.

2. CONTAINER INTEGRITY ANALYSIS SYSTEM

As shown in Figure 1, the proposed container integrity analysis system using the rear side image of the container consists of an input module that cuts out only the rear area of the container with resizing it to a certain size, a parts-zone partitioning module for dividing each component region, a module for analysing integrity parameters for each divided parts-zone, and a module for finally calculating the entire degree of integrity respectively.

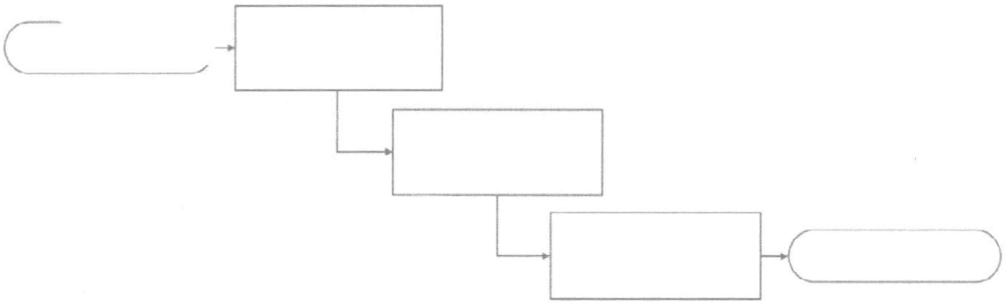

Figure 1. Block diagram of the proposed method

From the image of the container obtained by mobile devices, a square image is extracted by perspective transformation using four corner vertexes of the container area extracted with a manual or semi-automatic method as shown in Figure 2.

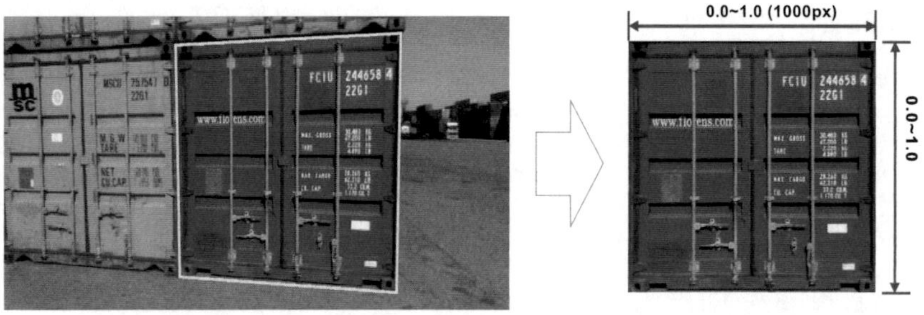

Figure 2. Rear side image extraction and size normalization

In order to minimize the influence between each accessory part in analysis, the input image is divided into parts-zone, each of them is analyzed separately, and the results are integrated to estimate the total degree of integrity. These processes will be described in more detail in the following sections.

2.1. Parts-zone partitioning

In analyzing the damage to the container through the general image processing techniques, various shapes of parts in the container may be mistaken for damage factors, which may affect the accuracy of the analysis. In order to solve this problem, the container image is divided into the areas of parts as analytic units named parts-zones to calculate the integrity parameters (contamination score, decrepitude score, locking score, etc.) by specialized algorithms for each areas. To do this, zone grids are first set up to divide the rear side image of the container into zone areas. The zone grids are lines to divide in which the relative positions are predefined based on a container standard specification template to distinguish each parts area in the size-normalized image as shown in the left of Figure 3. At this time, if the zone grids do not exactly match the container image due to the camera direction or distortion, an automatic adjustment function using the matching method of the LBP (Local Binary Pattern) mask may be performed to move the position of each zone grid within a certain range. After the zone has been determined by the zone grid, those are recombined to form the parts-zones as shown in Figure 3.Here, the parts-zones of door panel, door seal, and locking handle which are closely related to the life of the container, are selected as the parts-zone of interest for analysis.

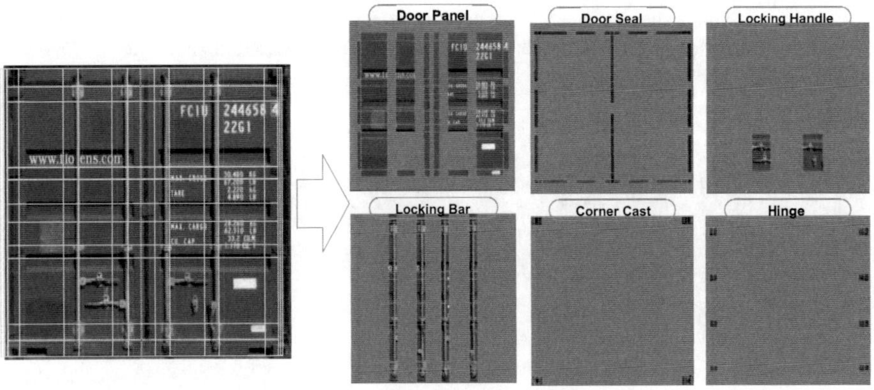

Figure 3. Parts-zone partitioning

2.2. Estimation of integrity parameters in each parts-zone

For each parts-zone, the integrity parameters are defined according to the major damage types expected in it. The contamination score for the door panel, the decrepitude score for the door seal, and the locking score for the locking handle are defined and calculated as these parameters. To compute these values, we applied a specific procedure for each parts-zone by combining color, edge, and frequency analysis methods properly as follows.

First, the contamination analysis procedure of the door panel is shown in Figure 4. The hue channel is extracted from the parts-zone of the door panel and the standard deviations of each zone in the parts-zone are calculated. If this value is above a certain threshold, it can be said that the zone contains contamination. The total contamination score of the door panel is calculated as a weighted average of the standard deviations multiplied by the size of each zone. Because various labels such as symbols or characters may act as noise components, the label masking step which extracts and removes the label area using thresholding and morphological operations of the gray image is performed.

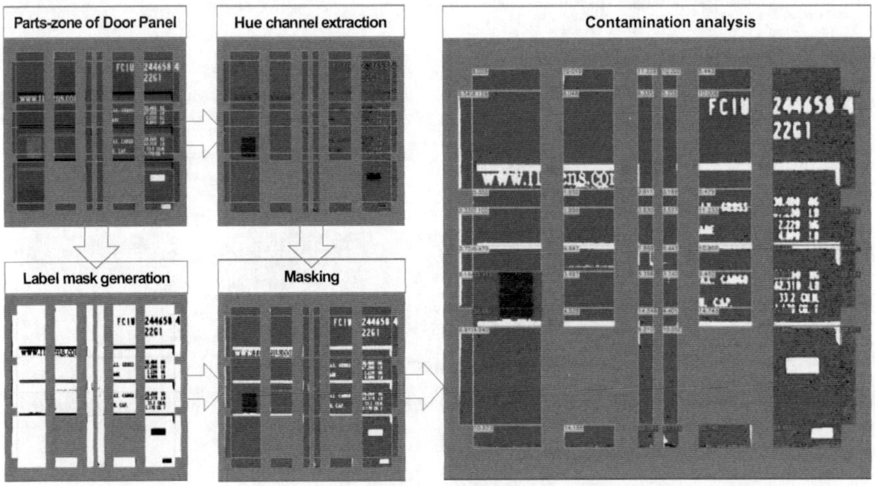

Figure 4. Integrity analysis for each parts-zone (I): Contamination score of door panel

Second, the decrepitude analysis procedure for the door seal as shown in Figure 5, uses the intensity channel extracted from the parts-zone image of the door seal. Through frequency analysis of the intensity image, the noise values of each zone constituting the parts-zone are obtained. The weighted average of the standard deviation with respect to the area size is calculated as the decrepitude score of the door seal.

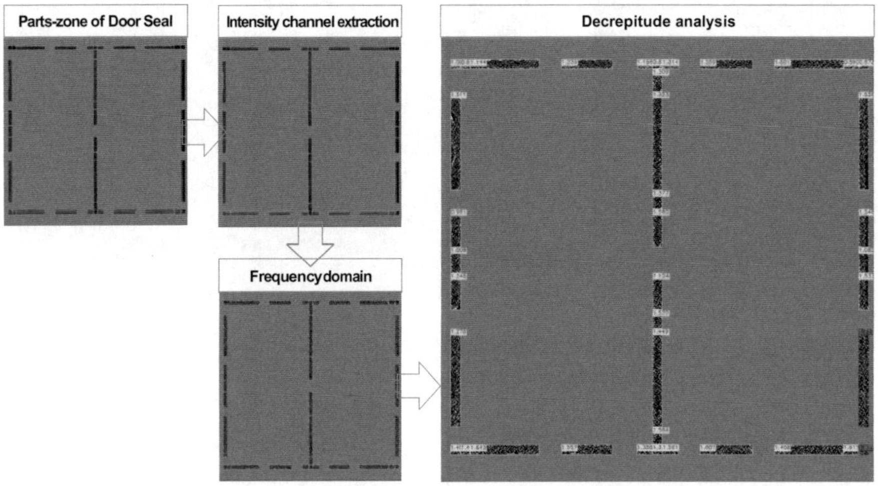

Figure 5. Integrity analysis for each parts-zone (II): Decrepitude score of door seal

Finally, Figure 6 shows the procedure for analyzing the locking score that indicates whether the locking handles are properly locked. If the Locking Handle is in the normal locked state, it is arranged in the horizontal direction. Therefore, we can obtain the locking score by extracting the horizontal edge components in the parts-zone of locking handle, performing appropriate noise removal, and calculating the average intensity of the edge.

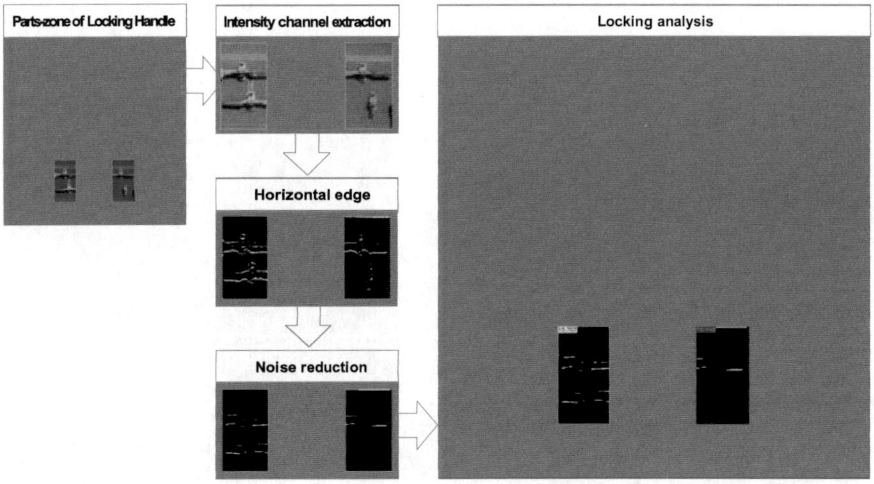

Figure 6. Integrity analysis for each parts-zone (III): Locking score of locking handle

2.3. Calculation of total degree of integrity

Through the procedures described above, result images showing the locations of the zones in which the damage exceeds a certain criterion are provided to the user, and the degree of integrity for the entire container is finally calculated by the weighted sum of the integrity parameter values from each parts-zone. If the degree of integrity does not meet a certain threshold, we can decide that this container is damaged enough to need repair.

3. EXPERIMENTAL RESULTS

Table 1 shows the results of the integrity analysis by proposed method for the six sample container images shown in Figure 7.

(a) ID-CON1 (b) ID-CON2 (c) ID-CON3 (d) ID-CON4 (e) ID-CON5

Figure 7. Sample images used in container integrity analysis

The contamination score of door panel (CDP), decrepitude score of door seal (DDS), and locking score of locking handle (LLH) as the integrity parameters for each rear side image of container had been calculated and the total degree of integrity (DOI) had been calculated by weighted sum of them. In some case, the CDP had been inaccurate due to the lighting conditions such as sunlight reflection on the door panel at the time of photographing, but the reasonable total DOI had been obtained by adjusting the weights.

Table 1. Results of integrity analysis with the sample images

Container ID	CON1	CON2	CON3	CON4	CON5
CDP	97.4	86.6	83.4	83.6	95.3
DDS	92.2	83.6	80.9	79.6	79.2
LLH	100.0	75.0	75.0	50.0	0.0
Total DOI	95.3	82.8	81.3	74.9	68.2

4. CONCLUSIONS

The container integrity analysis system proposed in this paper had divided the rear side image of the container photographed using a mobile camera into predefined parts-zones based on the container standard specification and applied an independent image processing algorithm for each parts-zone to measure the integrity parameters related to the container life. It then had calculated the degree of integrity for the entire container with these parameters.

The proposed system is inexpensive because it uses only the camera of the mobile device instead of additional expensive equipment. In addition, when run concurrently with a container management ledger computerization system, it is possible to provide more objective container integrity analysis results to stakeholders through photography alone without additional processes.

ACKNOWLEDGMENTS

This work was supported by Institute for Information & communications Technology Promotion (IITP) grant funded by the Korea government (MSIT) [No. R7119-16-1014, Development of the smart AEO (Authorized Economic Operator) automation platform based on standard IoT platform for import and export SMEs].

REFERENCES

Nakazawa K, Iwasaki I, Yamashita I (1995). Development of Damage Detection System for Container, Proceedings of the 1995 IEEE IECON 21st International Conference, 2, 1160-1163.

Huang GR, Su SF, Chen QJ, Lin YS (2013). Study on Content-Independent Feature Matching Systems for Container Images, 2013 International Conference on System Science and Engineering (ICSSE), 31-36.

LASE Industrielle Lasertechnik GmbH (2014). LaseCDI-GATE – Container Damage Inspection, Product data sheet.

Proc. 1st International Conference on Machine Learning and Data Engineering (iCMLDE2017)
20-22 Nov 2017, Sydney, Australia

High Efficiency Facial Expression Recognition based Active Semi-Supervised Learning

Kim Jin Woo[1], Minhaz Uddin Ahmed[1], Md Rezaul Bashar[2] and Phill Kyu Rhee[1]

[1] Department of Computer Engineering, Inha University, Incheon, South Korea

[2]Science,Technology and Management Crest, Sydney, Australia

Corresponding author's E-mail:pkrhee@inha.ac.kr

Abstract

Face expression recognition is an important area of computer vision research which has been extensively used for HCI (human computer interaction), identity confirmation, persons emotion recognition and different access control. It is difficult to recognize Human's facial expression in the real-world due to various circumstances, such as face dissimilarity in pose, ages, lighting condition, occlusion and face motion. Large amount of labeled and trained facial expression data can be one of the key idea. If the training dataset and test dataset collect from a particular person and environment the chances of getting higher accuracy is possible due to similar variation in facial expression. In this paper, we propose a method for gathering many facial expression data within an indoor environment and train them through ASSL(Active Semi-Supervised Learning) framework to gain high accuracy. We gather a large number of unlabeled facial expression data from intelligent technology Lab members of Inha University and BU-3DFE (Binghamton University 3D Facial Expression benchmark datasets. We train our initial model with the ASSL framework using deep learning network VGG(visual geometry group, University of Oxford) model. Our framework adopted MTCNN(Multi-task Cascaded Convolutional Networks) detector for face detection and we also modify the last two layer of VGG network for better performance. Repeat this entire process support us to get better performance improvement. Therefore using the ASSL method, we gain better performance and higher accuracy with less labor force. Our experimental result shows the high efficiency with various training data.

Keywords: Expression recognition, emotion classification, face detection, convolutional neural network, face recognition.

1. INTRODUCTION

As technology advances, machines become increasingly important in modern society. Especially, in the area of simple labor, many machines are replacing human labor, and as machine learning technology develops, more and more complicated tasks are also becoming the domain of machines. But there are still some tasks where machines cannot perform well. The reason is because there is a crucial and large barrier of emotion between man and machine. It is necessary to go beyond the high obstacle for smooth communication between human and computer.

The emotion of a person varies where facial expressions, voices, tone, gestures, words and phrases are also important factors. Therefore, machines need to observe and learn these factors in order to grasp human emotions. A person tries to understand other person's emotion by judge the surrounding situation and face expression. Whether it is for a short period of time but it is natural for every human being. But this is not a natural ability. It is because the human brain has learned how to grasp the emotions of the human beings little by little. People live in various situation and learn such process naturally whether the person is angry or the person who glares the eyes is unpleasant or the person who laughs wide open.

Yet machines do not follow the human brain's abilities. Early psychologist A. Mehrabian (1968) found that 7% of the person's emotions expressed in the language, 38% of the sentences delivered through the speech, and 55% of the emotions conveyed by the human facial expressions. Therefore, facial expressions can be very valuable information that can examine human behavior, consciousness, and mental activity. Thus, research has long been carried out to recognize facial expressions more automatically, efficiently, and accurately by Xiaoming Zhao, Shiqing Zhang (2016).

The research on facial expression recognition can be divided into two parts a) frontal facial expression recognition and b) non-frontal facial expression recognition. Among them, facial expression recognition is actively studied and has a high recognition rate. Especially since the convolution neural network is applied, the recognition rate goes up by Changxing Ding (2015).We propose a new method to obtain better recognition rate with less effort even change of environment does not affect the performance.

We organize this paper as follows. In Section 2, we explain the background information of incremental active semi-supervised learning with facial expression recognition system. In Section 3, we describe the facial expression recognition methods. In Section 4 and Section 5, we have explained details about the dataset and experimental environment with results. Finally, in section 6 we discuss the conclusions drawn from the experiments and future work.

2. BACKGROUND

A special kind of neural network known as CNN (convolution neural networks) has been widely used to solve face recognition problem with many different computer vision problems by Alex Krizhevsky, Ilya Sutskever and Geoffrey E. Hinton (2012). When training new data through CNN, overfitting problems are likely to occur. Overfitting problem such as where the learned model classifies well in the similar dataset but cannot classify well in new dataset. In addition over fitting has various drawbacks such as excessive training and tuning etc. However, the simplest solution is to create a training model using diverse and large amounts of training data so that new data can be classified well.

There are different types of face expression data coming out every day, it is almost impossible to perform supervise learning by labeling all these things. It requires too much labor and cost. Labeling a large volume of data is time consuming and tiring work for human but it is convenient for the computer due to repetitive work. On the other hand when overfitting problem occurs computer cannot tackle correctly. In order to solve this problem, a better learning approach is labeling new data using active semi-supervised learning method.

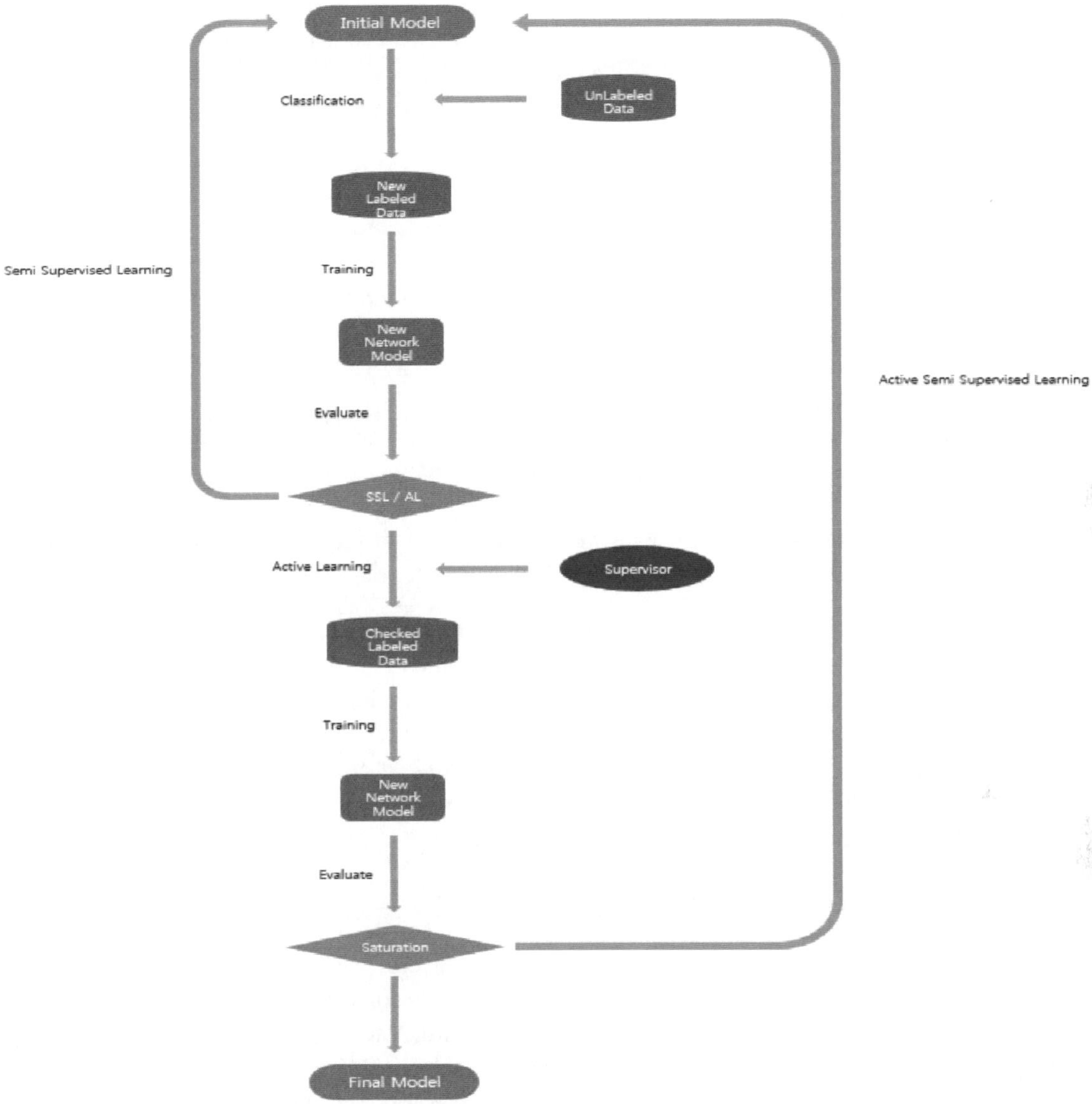

Figure 1: active semi-supervised learning model

Figure 1. shows the overall design of the propose system. First of all, we train the BU3DFE database to create a pre-learning model with excellent vector extraction capabilities because 2D-based analysis is difficult to handle large pose variation and slight facial expression recognition by Lijun Yin, Xiaochen Chen, Yi Sun, Tony Worm and Michael Reale (2008). Next, an initial learning model is created through active learning based on a pre-learning model. Subsequently, we use the semi-supervised learning with active learning where active learning reduces the wrong face identification and gradually improve the performance of the learning model. If the performance is lower than that of the previous training result, the training data is discarded and the next training data is used to perform the training again.

Active learning is performed when the result of the learning is not improved smoothly. We compare threshold value with estimated labeled score from face detector generate through semi-supervised learning. If the labeled score is below threshold value we do not consider for the active learning process. A confidence value is between 0 and 1 whereas threshold value is 0.9 because lower threshold value biases the outcome in our experiment.

3. METHODS

In facial expression recognition research, seven facial expressions are considered such as angry, disgust, fear, happy, sadness, surprise and neutral. However, different level of intensity increases the variation of facial expressions let alone these seven expressions. American psychologist Paul Ekman (1993) proposed the concept of basic six emotions, which depicts the six facets of anger, dislike, fear, happiness, sadness, and amazement by H. Rowley, S. Baluja, and T. Kanade(1995). In fact, the expression of a person is not only affect by emotion but also various factors such as appearance, age, sex, and culture. For example, in case of the West and the East, the expression is same but the emotion is different. Western people usually express joy as it is, but in case of oriental considers it humbly. This is a result of learning by a cultural difference between western and oriental. Also, young people express their emotion differently compared to elder people and handsome people are more confident than those who are unattractive.

The overall procedure of facial expression recognition is shown in Fig.1. First of all, a face image is capture by webcam as input then normalization and segmentation is performed. Normalization removes the changes of illumination or face location so that the facial image has uniform characteristics. Partitioning is a technique for extracting only a part of a given image including a necessary facial image. This preprocessing process has a great effect on accuracy because it plays a role in remove noise which obstructs recognition by B. Fasel, Juergen Luettin (2012), Andre Teixeira Lopes, Edilson de Aguiar, Alberto F. De Souza, Thiago Oliverira-Santos(2017).

We can easily distinguish the facial expression at a glance by looking at the face of an angry expression, a smiling expression or a surprised expression. The reason for this is that we can quickly identify the features of the raised eyebrows, tight lips, and wide mouths. The characteristic vector is the symbol of the characteristic of each facial expression so that the machine can recognize it like a person. Therefore, in order to recognize facial expressions, it is necessary to extract feature vectors that can be clearly distinguished by applying feature extraction techniques such as Deformation Extraction or Motion Extraction for the result of preprocessing adopted by B. Fasela and Juergen Luettin (2003). The transformation extraction is to express an image or a model including a geometric transformation of a facial expression as a feature vector, and an operation extraction is a feature vector expressing a change amount of a facial expression according to the passage of time.

The extracted feature vectors are classified by modifying the layer of VGG network so that they could be identify through facial feature representation. After that, classifiers are used to classify facial feature expressions according to the input labels. After that, the classifier has a value that can distinguish the facial feature representation for each label. The classifier classifies the extracted feature vectors when receiving a new face image with this value, where 1 denotes an angry expression, 2 denotes a disgusting expression, 3 denotes a fearful expression, 4 denotes a smiling expression, 5 denotes a sad expression, 6 denotes surprised facial expression, and 7 is expressionless.

Convolutional neural networks are one of the new machine learning schemes that have received attention in recent years due to better performance in resolving computer vision problems. The convolution neural network extracts and classifies the feature vectors according to the network structure determined by the user. CNN extracts appropriate feature vectors from each image and classifies them into the label.VGG very deep 16 network was used for learning facial expression recognition. The network adds a very small (3x3) convolution filter to the existing VGG network, resulting in better performance for large image recognition by Ken Chatfield, Karen Simonyan, Andrea Vedaldi, Andrew Zisserman (2014), Karen Simonyan, Andrew Zisserman (2015).

4. DATA SET

The BU3DFE database is used to create a pre-learning model for this experiment. The database includes diverse races, including West, East, East-Asian, Middle-east Asian, Indian, Hispanic Latino, and others. The BU3DFE dataset provides facial expressions for 100 people, ranging from 56% women and 44% men between the ages of 18 and 70. The types of facial expressions provided by Paul Ekman (1993) basic 6 emotions are added to the expression, there are seven labels of anger, disgust, fear, happiness, sadness, surprise and expressionless. There are a number of emotion levels by Lijun Yin, Xiaochen Chen, Yi Sun, Tony Worm, Michael Reale (2008). We selected 700 images (7 x 100) of images with the strongest emotion level of the frontal pose and conducted preliminary learning.

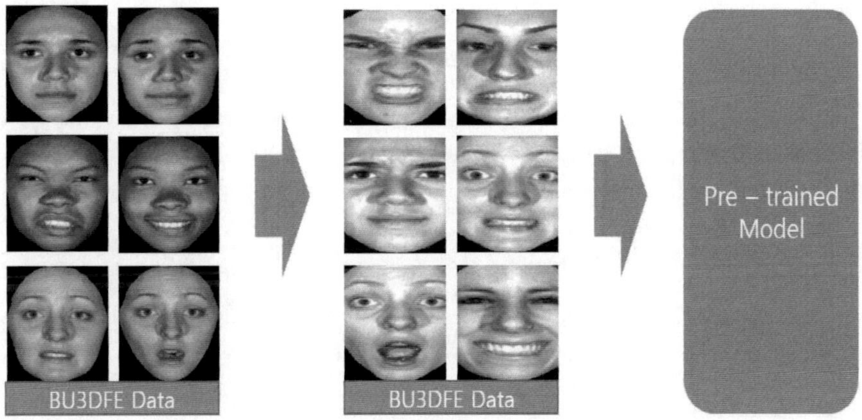

Figure 2: Pre-trained model

Figure 2. shows the BU3DFE dataset that use to transform pre-trained model before image preprocessing technique such as noise elimination and intensity standardization applied to increase the facial expression recognition rate. Then, the modified data is learned by the VGG-based deep learning network model, and an initial learning model for semi-supervised active learning method is created. In order to measure the performance, five experimental data consisting of 35 images is prepared separately for each emotion. Through this, the performance is continuously measured while learning through the active learning method, and the average is obtained. We have continuously measured the performance change of the previous learning model for the new environment by changing the experimental environment whenever performance converges in the process.

5. EXPERIMENTAL RESULT

Our experiments use publicly available VGG net that has 5 convolutional layers and 3 fully connected layers. This networks are pre-trained ILSVRC-2013 dataset by A. Berg, J. Deng, and L. Fei-Fei. (2010), which includes 1.2M training images , labeled into 1000 classes. For the detection system we use MTCNN by K. Zhang and Z. Zhang and Z. Li and Y. Qiao Joint (2016) that is convolutional neural network based face detector and ASSL frameworks is implemented on the popular deep learning library matconvnet and Caffe by Yangqing Jia,Evan Shelhamer, Jeff Donahue, Sergey Karayev, Jonathan Long, Ross Girshick Sergio Guadarrama and Trevor Darrel (2014). All implementations are on a single server with cuDNN S. Chetlur and C. Woolley (2014) and a single NVIDIA GeForce GTX 970 graphics card with operating system ubuntu 14.04.
We decide the order for six different environments. We use the initial learning model to measure the initial performance in each environment. Then we measure initial performance in ascending order to start learning from the environment with the lowest initial performance. Table 1 shows the order of data.

Table 1 Initial performance and order of data for 6 environments

Distinction	Front	Left	Right	Glasses	Hat	Hat & Glasses	Mask
Initial performance	71.4	51.4	85.7	34.2	94.2	65.7	31.4
Order	5	3	6	2	7	4	1

The reason for setting the learning sequence in this way, because of active learning starts from an environment with a high initial performance. If we start active semi-supervised learning from high initial performance environment than semi-supervised learning predict incorrect labels with high confidence. Therefore, to minimize this process, we gradually attempted to learn from the environment most similar to the data used to construct the initial learning model.

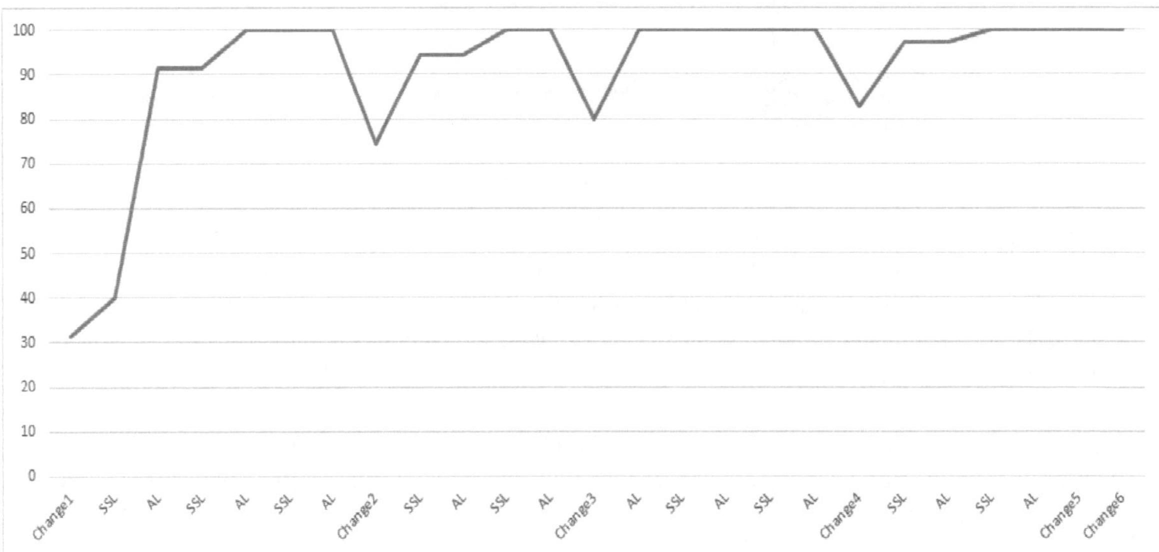

Figure 3: Performance of High-Efficiency Adaptive Facial Expression Recognition System by Active Learning

Experimental results show that the convergence of the initial performance decreases as the environment changes. This is because the learning of various environment data is automatically learned through active learning. Figure 3. shows the initial performance changes due to environmental changes.

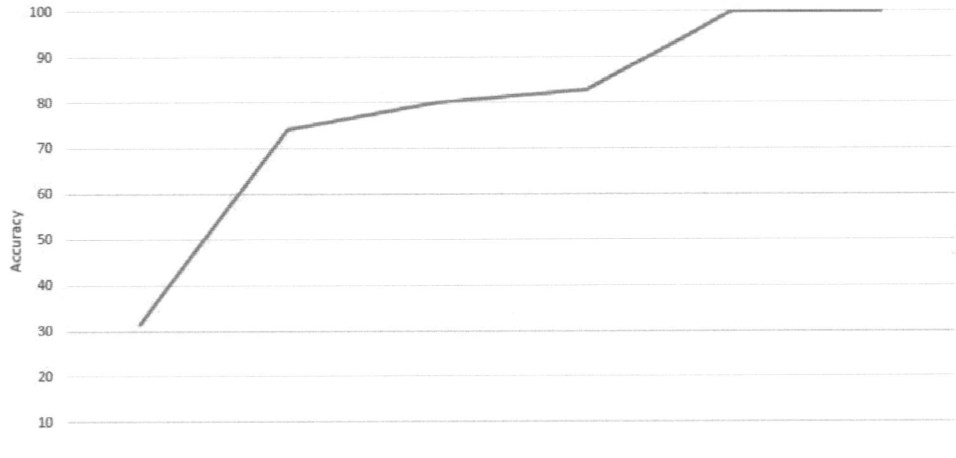

Figure 4: Initial recognition rate increase in graph due to environmental change

This experiment also requires much less manpower than labeling all the collected data. As shown in Figure 4, it can be seen that the labor required for manual labeling is 3.5 times more than the active semi-supervised learning method.

6. CONCLUSION

In this paper, we propose a semi-supervised learning framework for face expression recognition that can adapt to various environmental changes with the low labor force. In the proposed experiment, the undesirable side of this framework is label with wrong expression with high reliability which caused the problem for performance improvement due to the wrong learning result. However, by adjusting the good dataset order by comparing the initial performance, this problem can be solved by excluding the data from the experiment when the performance is not good. The proposed system is expected to help a lot when requires a large amount of training data in various environments. Our future research direction is to find a way to train sequence of face expression images with fewer images. In addition, by reducing the number of erroneous labels with high reliability can minimizing the discarded learning data and maximize efficiency.

7. ACKNOWLEDGEMENTS

This work was supported by the ICT R&D program of MSIP/IITP. [2017-0-00543], Development of Precise Positioning Technology for the Enhancement of Pedestrian's Position/Spatial Cognition and Sports Competition Analysis].The GPUs used in this research were generously donated by NVIDIA.

REFERENCES

A. Mehrabian (1968). Communication without words, Psychol. Today, Vol. 2, pp. 53_5

Xiaoming Zhao, Shiqing Zhang, (2016). A Review on Facial Expression Recognition - Feature Extraction and Classification, IETE Technical Review.

Changxing Ding (2015). Robust Face Recognition Via Multimodal Deep Face Representation,IEEE TRANSACTIONS ON MULTIMEDIA, VOL. 17, NO. 11.

B. Fasel, Juergen Luettin (2012). Automatic facial expression analysis: a survey, International Journal of Computer Science & Engineering Survey (IJCSES), Vol.3, No.6.

H. Rowley, S. Baluja, T. Kanade (1995). Human face detection in visual scenes. Technical Report CMU-CS-95-158R, School of Computer Science, Carnegie Mellon University.

Andre Teixeira Lopes, Edilson de Aguiar, Thiago Oliveira-Santos (2015). A Facial Expression Recognition System Using Convolutional Networks, 28th SIBGRAPI Conference

Ken Chatfield, Karen Simonyan, Andrea Vedaldi, Andrew Zisserman (2014). Return of the Devil in the Details: Delving Deep into Convolutional Nets, arXiv:1405.3531v4 [cs.CV].

Karen Simonyan, Andrew Zisserman (2015).Very deep convolutional networks for large-scale image recognition, arXiv:1409.1556v6 [cs.CV].

Cohn, D., Ghahramani, Z., I. Jordan M (1999). Active learning with statistical models. Journal of Artificial Intelligence Research, Vol.4, pp. 129-145,

Riccardi, G. and Hankkani-Tur, D (2005). Active learning: theory and applications to automatic speech recognition, IEEE Transactions on Speech and Audio Processing, Vol.13, No.4, pp. 504-511

Andre Teixeira Lopes, Edilson de Aguiar, Alberto F. De Souza, Thiago Oliverira-Santos (2017). Facial expression recognition with Convolutional Neural Networks: Coping with few data and the training sample order, Pattern Recognition Volume 61, Pages 610–628.

Gil-Jin Jang, Ahra Jo, Jeong-Sik Park, Yong-Ho Seo (2014). Video-based Facial Emotion Recognition using Active Shape Models and StatisticalPattern Recognizers, JIIBC, VOL. 14 NO. 3, pp.139-146.

Sinno Jialin Pan, Qiang Yang (2010). A Survey on Transfer Learning, IEEE Trans. Knowl. Data Eng. 22, 1345-1359

Lijun Yin, Xiaochen Chen, Yi Sun, Tony Worm, Michael Reale (2008). A High-Resolution 3D Dynamic Facial Expression Database, The 8th International Conference on Automatic Face and Gesture Recognition.

K. Zhang and Z. Zhang and Z. Li and Y. Qiao Joint (2016). Face Detection and Alignment Using Multitask Cascaded Convolutional Networks, IEEE Signal Processing Letters

Steve Lawrence, C. Lee Giles, Ah Chung Tsoi, and Andrew D. Back (1997). Face Recognition: A Convolutional Neural-Network Approach, IEEE Transactions on Neural Network

Lijun Yin; Xiaochen Chen; Yi Sun; Tony Worm; Michael Reale (2008). A High-Resolution 3D Dynamic Facial Expression Database" The 8th International Conference on Automatic Face and Gesture

Ekman, Paul (1993). Facial expression and emotion. American Psychologist, Vol 48(4), 384-392.

B. Fasela and Juergen Luettin (2003). Automatic facial expression analysis: a survey, Elsevier, pattern recognition, 36 (2003) 259 – 275

A. Berg, J. Deng, and L. Fei-Fei. (2010). Large scale visual recognition challenge (ILSVRC), URL http://www.image-net.org/challenges/LSVRC/2010/.

Yangqing Jia,Evan Shelhamer, Jeff Donahue, Sergey Karayev, Jonathan Long, Ross Girshick Sergio Guadarrama and Trevor Darrel (2014). Caffe: Convolutional Architecture for Fast Feature Embedding, ACM international conference on Multimedia

S. Chetlur and C. Woolley (2014). cuDNN: Efficient Primitives for Deep Learning, arXiv Prepr. arXiv. 1–9

Alex Krizhevsky, Ilya Sutskever and Geoffrey E. Hinton (2012) ImageNet Classification with Deep Convolutional Neural Networks, Advances in Neural Information Processing Systems 25

Proc. 1st International Conference on Machine Learning and Data Engineering (iCMLDE2017) 20-22 Nov 2017, Sydney, Australia

The Optimization of Distributed Computed SVM by KNN

Lizelong[1] and Yangcheng[2]

[1]Zhejiang university, Hangzhou, China

[2]Zhejiang university city college, Hangzhou, China

Corresponding author's E-mail: yangc@zucc.edu.cn

Abstract

Support Vector Machine is one of the most popular classification algorithms that is based on statistics learning. SVM has many advantages such as over-fitting and disaster effect. One of the challenges for SVM is it cannot fit for large samples. When we train a model by isolated SVM for a large set of data, the execution time to train the model is extremely high and often unacceptable.

To solve this issue we propose a Cascade SVM algorithm, which is a parallel algorithm with a good effect that can sharply reduce the cost of training time. We split the large-scale dataset into several small-scale datasets and concurrently iterate those clusters in order to increase its train speed. Therefore, the cost is further reduced. However, it produces lower accuracy compared to isolated SVM. To increase the accuracy, we use k nearest neighbour (KNN) algorithm. If the distance between a data point and hyperplane is less than the threshold value when SVM is predicting the test data, we accept the prediction result. Otherwise we consider randomly combine the KNN's result. In this way, we train our model efficiently. Therefore, our proposed model will increase the accuracy and will reduce the prediction speed obviously.

Keywords: Cascade SVM, Parallelization, KNN

1. INTRODUCTION

As the progress of time, the speed of data growth has increased. In such an era of big data, large samples of data can't be processed by the conventional classification algorithms in an acceptable scope of time, so we need to improve these algorithms.

Support Vector Machine (SVM) is one of the most popular classification algorithms mentioned by P. H. Chen, C. J. Lin, B. sSchölkopf (2005) and C. J. C. Burges (1998). And it is based on statistics learning by V. N. Vapnik (1999). SVM has many advantages such as over-fitting and disaster effect. But one of the challenges for SVM is that it cannot fit for large samples. And k nearest neighbor (KNN), supposed by TM Cover and PE Hart (1967), can use a shorter time to train samples with high accuracy. However its prediction calculation costs a large amount of time, and it is prone to classify the data into wrong categories when it has few train samples.

Therefore, this paper proposes a new parallel SVM algorithm which is based on the algorithm of Cascade SVM from Hans Peter Graf (2005). It has high concurrency than before, and by using KNN to optimize the result, the parallel can avoid the accuracy reduced problem about the parallel SVM algorithm.

2. RELATED WORK

2.1. SVM algorithm

Support Vector Machine algorithm is one of the powerful classification algorithms. The basic idea of SVM is to find an optimal separating hyperplane by Peter Harrington (2012), it not only can split different kinds of data, but it can also ensure that the interval is the largest. Figures 1 and 2 show that both of the hyperplanes can split different kinds of data, but the min interval from sample point to hyperplane, shown in Figure 2, is the largest. So, this hyperplane is the optimal separating hyperplane, and the sample points at the separated boundary are called support vectors.

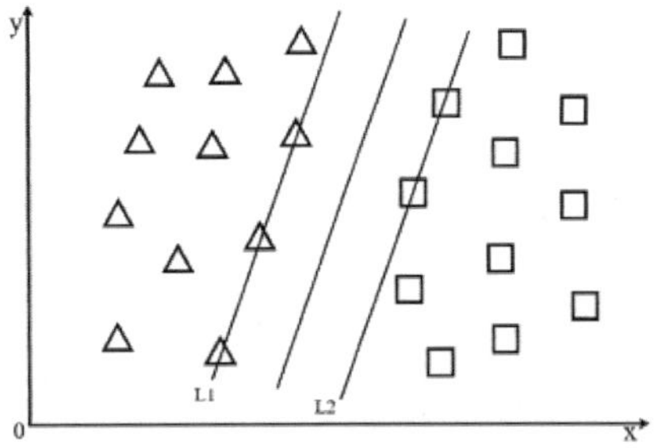

Figure 1. Optimal Hyperplane with Largest Interval

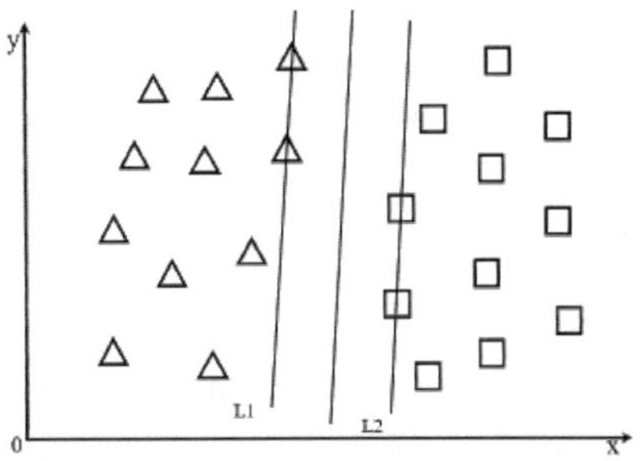

Figure 2. Normal Hyperplane without Largest Interval

Now we assume the training sample is x_i, the prediction sample is y_i, and $w{\cdot}x+b=0$ is the classify hyperplane. If x_i meet $|w{\cdot}x+b|\geq 1$, then x_i is support vector. Optimal hyperplane must meet the equation (1)

$$y_i[(w \cdot x_i) + b] \geq 1 \tag{1}$$

Then the classify interval is:

$$\mathrm{d} = \min_{\{x_i|y_i=1\}} \frac{(w \cdot x_i + b)}{||w||} - \min_{\{x_i|y_i=-1\}} \frac{(w \cdot x_i + b)}{||w||} = \frac{2}{||w||} \tag{2}$$

To get the max interval, we should make ||w|| min, so we can compute the min of equation (3) with constraint condition $y_i\,[(w{\cdot}x_i)+b] \geq 1$

$$\varphi(w) = \frac{\|w\|^2}{2} = \frac{(w \cdot w)}{2} \tag{3}$$

Therefore, when the sample is linearly inseparable, we should use a non-negative slack variable Gamma and a positive penalty parameter C. If the sample is nonlinearity, we should use kernel function to transform the sample into a high-dimensional linearly separable characteristic space. After that, we can find the optimal classify hyperplane.

2.2. KNN algorithm

The main idea of K Nearest Neighbor (KNN) algorithm is to compute the distance between test samples and train samples and finding the nearest k neighbors. One of the distance is Euclidean distance, just as equation (4).

$$\mathbf{d}(\mathbf{x}, \mathbf{y}) = \sqrt{\sum_{i=1}^{n}(x_i - y_i)^2} \tag{4}$$

In the KNN algorithm, all the neighbors are correctly classified. The method of classifying only bases on the most adjacent one or a few sample's category. When the train samples are not balances, such as when a category of a sample's size is very large and other types of a sample size is very small, it can lead to the large size classify result when we input a new test sample.

2.3. Cascade SVM algorithm

The cascade SVM is an optimized SVM which uses parallel for accelerating SVMs. Because the support vectors are just a small part of the training sample, we can split the large dataset into a lot of sub datasets. Then we can train the sub dataset to improve the whole training speed. The goal of SVM training is to eliminate the redundancy support vectors. When all training processes eliminate all the redundancy support vectors surrounding sub datasets, we can finally get the global support vectors.

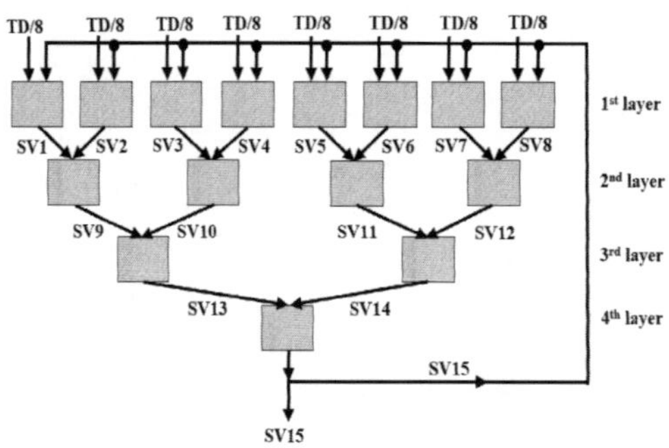

Figure 3. Structure of Cascade SVM

Based on the thought of eliminating the non-support vectors in every iteration, Figure 2 shows the TD and the arrows between layers which represent the training samples and process of SVM training respectively. In the first layer, we split the original sample into 8 TDs, then we used the Sequential Minimal Optimization (SMO) algorithm. The SMO algorithm, published by JC Platt (1999), can break down the original, complicated and large optimal problems into simpler ones, to train TDs and get support vectors. We then merged the support vectors into 4 parts of SV as the next layer's TDs. In the second layer, we repeat the first layer operation. Repeating this process we finally get the final training sample TD15 in this iteration. We then judge whether the TD15 can fit the threshold value. If it does not fit, we should repeat the above-mentioned process until it fits the requirement.

Through the above-mentioned approach, cascade SVM can sharply increase the training speed and get an effective model.

3. METHODS

Cascade SVM uses distributed systems to improve training speed, and set the threshold for iteration to improve the prediction accuracy. Therefore it can indeed solve the low speed of isolated SVMs handling the big dataset to some degree.

However it still has some serious problems:

1) It is done by binary merge, so the concurrent thread number will dramatically decrease.
2) As the algorithm runs, because of the invalid support vectors that are removed, it cannot effectively reduce the support vectors like the begin phase. So, the accuracy will almost stop to increase.
3) In the later algorithm phase, the number of merged support vectors will be larger, so the training speed will be slower and slower.

Therefore, we use a new method to handle these problems. We give up the iteration to get higher efficiency and increase the accuracy. The algorithm is as follows:

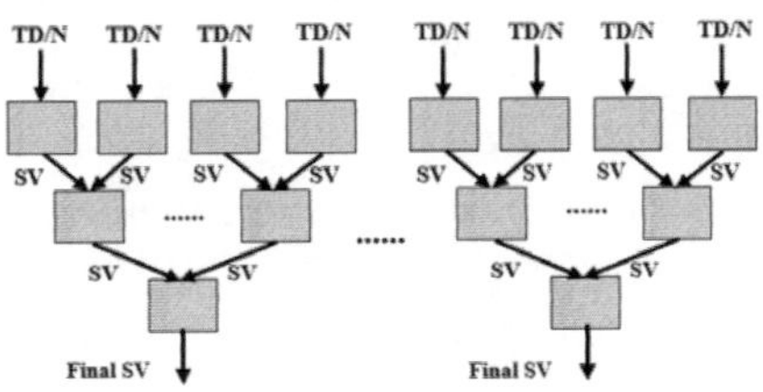

Figure 4. Structure of Distributed Computed SVM by KNN

Before the training phase, we shuffled the original sample in order to train different sizes of the subset samples and get their accuracy. This provided an appropriate block size named N. We set the threshold value P (less than or equal to the computer amounts of distributed system) to improve the train speed.

In the training phase, we split the original sample into N parts and used SMO algorithm to train every sub dataset. After that we binary merged the support vectors obtained from training. Finally, we looped the process until remaining P support vectors blocks. At this moment, the P computers saved different support vector blocks respectively.

In the prediction phase, we sent the sample for prediction to the P computers and got the P prediction answers. Then the P computers sent results to the center computer to merge the results. We set a threshold value R which can be chosen from 1/2 to 3/4. For the same sample point, if the number of common result were more than R, we accepted the result. If not, we used KNN to predict the sample point. If the result obtained by KNN existed in the SVM results, we also accepted it. Otherwise we randomly accepted the KNN results in a 50% ratio, with the other 50% ratio split into SVM result respectively.

4. EXPERIMENTS

4.1. Experiment Environment

Because there was not enough hardware, we used a single computer which has an Intel i7 6700HQ CPU, 16GB memory and 1TB hard disk.

For the software environment, we used Anaconda 4.4 which is based on Python 3.6, and we used Scikit-learning 0.18.1 which is a Python machine learning package and integrates the LibSVM proposed by Chang Chih-Chung, Lin Chih-Jen and which has reference site link is: http://www.csie.ntu.edu.tw/~cjlin/ (2017-10-10).

The Experimental data is the Covtype dataset provided by UCI Machine Learning Repository with site link address: http://archive.ics.uci.edu/ml/machine-learning-databases/covtype/ (2017-10-10), the dataset has 7 class, 54 attribute and 581012 sample point.

4.2. Experiment Environment

When the experiment began, we shuffled the sample and split the sample into two parts. The part which had 90% sample points is the training sample, and the other part is the test sample. In the experiment, the K parameter for KNN is 5. For the SVM part, we used RBF kernel function. The C is 2.46 and the gamma is 0.66. Because the experiment hardware's CPU is 4 cores, we used 8 threads to simulate 8 computers and we set the parameter P on 8, R on 0.5 and the block size was 6000.

Due to the lack of environment hardware, we used multithreading to simulate the distributed system. In order to make the experimental results effective and comprehensive, we randomly extracted 5000, 10000, 25000, 50000, 75000, 100000 sub dataset from the original sample. We then compared the difference between the isolated SVM, Cascade SVM and the SVM we proposed.

Figure 5 and6 show the comparison speed and accuracy between the isolated SVM, Cascade SVM and the SVM we proposed.

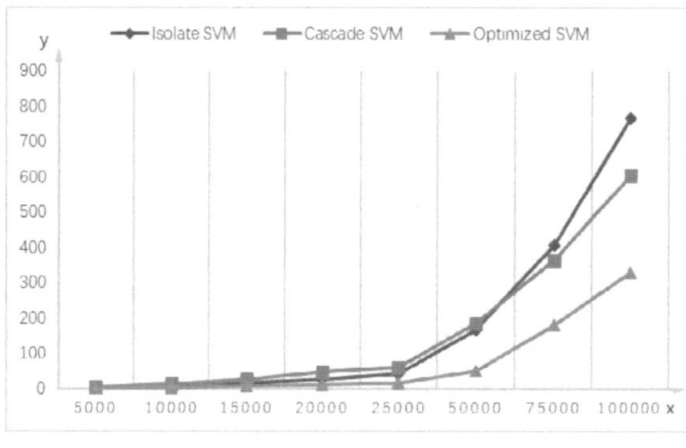

Figure 5. Train Speed Comparison

As can be seen in Figure 5, beginning threads start costs a lot of time and the cascade SVM costs more time than isolated SVM. As the sample size grows, the threads start cost becomes a small fraction of the sum time cost. Therefore, the cascade SVM costs less time than the isolate SVM at the end. However, the custom SVM time cost is always less than half of the isolated SVM time cost or the cascade SVM time cost. As a result, the custom SVM have good time performance. But due to the program running on a single computer, the concurrent cost about transfer files and inter-node communication cannot be shown in this figure.

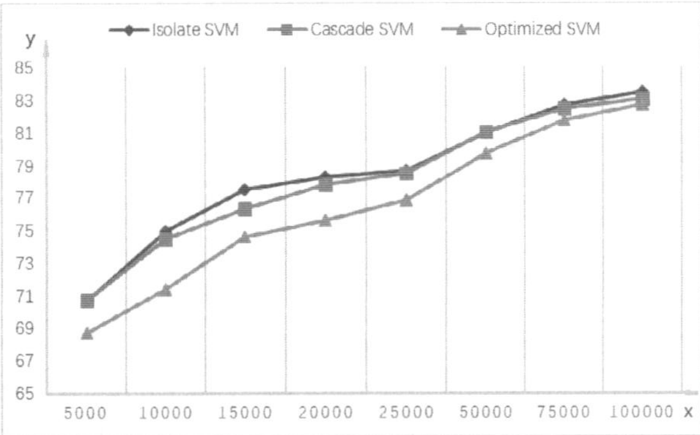

Figure 6. Prediction Accuracy Comparison

When the custom SVM deals with small train data scale, the split train blocks are too small to have a good prediction performance. Chart 1 shows that the custom SVM base on KNN algorithm prediction accuracy is similar to KNN's prediction accuracy. Figure 6 shows that when the train data scale grows, the prediction accuracy of KNN algorithm will be obviously increased. Therefore, the accuracy is similar to the isolated SVM and Cascade SVM in the end. The accuracy of custom SVM can be guaranteed.

Figure 5 and 6 shows that custom SVM has similar accuracy and costs less than half, compared with other SVM. Therefore custom SVM has good effect.

Chart 1. Prediction Accuracy Comparison between KNN and Custom SVM

	5000	10000	15000	20000	25000	50000	75000	100000
KNN	69.61	71.10	73.82	75.05	77.16	80.32	82.74	83.49
Custom SVM	68.73	71.38	74.61	75.59	76.86	79.73	81.76	82.73

From chart 2, we can see that most of the time uses SVM rather than KNN, and the ratio we use KNN algorithm will fall down. Therefore, the time cost can't grow significantly due to using KNN. And because the predicting functions run in different threads, the time cost in prediction phase is close to the isolated SVM.

Chart 2. Use KNN Ratio in Custom SVM

	5000	10000	15000	20000	25000	50000	75000	100000
Ratio (%)	19.8	17.1	17.1	15.5	16.08	12.38	11.41	9.85

As mentioned above, this algorithm has a more satisfactory effect than others.

5. CONCLUSIONS

In this work, we propose an optimized distributed computed SVM which is based on KNN. For Cascade SVM, this method aims at handling the decreasing concurrent and low efficiency in convergence and iterative phase. We eliminate the iterative phase to avoid low efficiency, and set a threshold value to increase the concurrent. Furthermore, we set an acceptance threshold value to identify the result and use KNN algorithm as the assistant algorithm to increase the accuracy further.

In comparison to other algorithms, experiments show that the method has better speed and similar accuracy.

Further study can use heuristic algorithm to reduce the dimension of sample and use better cascade architecture. And it can use Map-Reduce supposed by R Mmel (2008) via Hadoop, site by link address: http://hadoop.apache.org/ (2017-10-10), for more convenient use.

6. ACKNOWLEDGMENTS

This work was supported by Zhejiang University Research Grant in 2017.

7. REFERENCE

P. H. Chen, C. J. Lin, and B. Schölkopf (2005). A tutorial on ν-support vector machines, Appl. Stoch. Models. Bus. Ind. 2005, 21, PP.111-136.

C. J. C. Burges (1998). A tutorial on Support Vector Machines for pattern recognition, Data Min. Knowl. Discov. 1998, 2, PP.121-167.

V. N. Vapnik (1999). An overview of statistical learning theory, IEEE Trans. Neural Netw. 1999, 10, PP.988-999.

TM Cover, PE Hart (1967). Nearest neighbor pattern classification[J]. IEEE Transactions on Information Theory, 1967, 13(1), PP. 21-27.

Hans Peter Graf, Eric Cosatto, Leon Bottou (2005). Parallel Support Vector Machines: The Caseade SVM

Peter Harrington (2012). Machine Learning in Action, Manning Pubns Co, 2012

Vladimir N. Vapnik (1998). Statistical Learning Theory, Wiley-Interscience, 1998

JC Platt (1999). Fast Training of support Vector Machines Using Sequential Minimal Optimization[M] MIT Press, PP.185-208

Scikit-learn. http://scikit-learn.org/stable/

Chang Chih-Chung (2011), Lin Chih-Jen. LIBSVM: a Library for Support Vector Machine[J]. ACM, 2011(2), PP.1-27

Libsvm. http://www.csie.ntu.edu.tw/~cjlin/

UCI Machine Learning Repository, Covertype Data Set. http://archive.ics.uci.edu/ml/machine-learning-databases/covtype/

R Mmel (2008). Google's MapReduce programming model – Revisited. Science of Computer Programming, 2008, 70 (1), PP.1-30

Hadoop. http://hadoop.apache.org/

Proc. 1st International Conference on Machine Learning and Data Engineering (iCMLDE2017)
20-22 Nov 2017, Sydney, Australia

A Deep Learning based approach for Human Action Recognition

Minhaz Uddin Ahmed[1], Kim Jin Woo[1], Miyoung Nam[1], Md Rezaul Bashar[2] and Phill Kyu Rhee[1]

[1] Department of Computer Engineering, Inha University, Incheon, South Korea
[2]Science, Technology and Management Crest, Sydney, Australia

Corresponding author's E-mail:pkrhee@inha.ac.kr

Abstract

Action recognition is gaining importance due to its numerous uses in security assistance, video analysis, and surveillance application. Action recognition from an image and online video is challenging works due to the complex background, object scale, variation of pose and miss interpretation of action. In this paper, we proposed a Deep learning based method for human action recognition using Convolutional Neural Network (CNN).The CNN use jointly learning feature transformation which classifies the object in image optimally and provides better performance on the other hand traditional manually designed features cannot perform well. Deep learning based approach exploits GPU as well as computational resources and produces a competent performance on real time videos and intelligent technology Lab dataset of Inha University. In our experiment, we use different dataset images for training such as simple and noisy and cluttered environment. We have trained the intelligent technology Lab dataset by pre-trained network (VGG16 network by the University of Oxford) model. We have found cropped ROI with higher iteration has better action recognition than large image while performing test owing to the less noisy background. Total five activities were considered due to the limited number of labeled training examples. We also integrated an active semi-supervised learning (ASSL) method which helps in improving the overall human action recognition and achieved average accuracy of 96%. For testing, we recorded videos which contain phoning, taking photo, using computer, jumping and reading activities. Our approaches significantly achieved higher MAP on intelligent technology Lab dataset.

Keywords: Action Recognition, Convolutional Neural Network, Deep Learning, Transfer Learning.

1. INTRODUCTION

Human action recognition is an important problem because through this we can identify the real life person interaction, person movement and particular task carry out by a person. Still picture or video contains significant information about human actions. Interpretation and understanding of these actions from images helps to resolve real life problems such as surveillance video analysis for safety, elder patient monitoring, sports video analysis, detection of abnormal activities or irregular behavior, human computer interaction and so on. All these applications require efficient action recognition from video. Due to complex environment such as crowded place (e.g market, bus station), pose dissimilarity (e.g running and walking), person to person interaction (e.g playing), person to object interaction(e.g using phone or camera), low image quality , occlusion, noisy background, poor lighting condition and many other issues in the real time make it difficult to recognize human actions.

In order to solve this problem, we propose a Visual Geometry Group of Oxford University (VGG) network based deep learning model to recognize the human action with integrated Active semi-supervised Learning (ASSL) framework.
In recent years, Deep Convolutional Neural Network has been extensively used for computer vision applications especially human action detection by C. Feichtenhofer, A. Pinz, A. Zisserman (2016), G. Gkioxari, R. Girshick, and J. Malik (2015), H. Mahmudul, and Amit K. Roy-Chowdhury (2015) because of higher accuracy and appropriate application to real life problems. Over the last few years object recognition method become more developed than previous approaches such as the Deformable Part Model(DPM) by P. F. Felzenszwalb, R. B. Girshick, D. A. McAllester, and D. Ramanan (2010) , AlexNet by A. Krizhevsky, I. Sutskever, and G. Hinton (2012), VGGNet by Karen Simonyan, Andrea Vedaldi ,Andrew Zisserman (2013). We deploy VGGNet in our research work due to its promising results. Some other popular methods are based on a probabilistic model by Lei Zhang, Zhi Zeng, and Qiang Ji (2011) which works well among simple actions in a favorable environment with HMM by Hanju Li ,Yang Yi , Xiaoxing Li , Zixin Guo (2013) method. However, very complex action in a dynamic environment often impossible to detect correct action such as brushing hair and using the cell phone or fight and hug in an occluded environment.

In the next section we analysis the literature of action recognition in still images and videos. We describe our approach in section 3 action recognition based on transfer learning. In section 4 we describe experiments carry out based on the Intelligent Technology Lab action dataset. In section 5 we explained our experiment result and contribution.

2. RELATED WORK

Action classifications, activity recognition, human behavior analysis in image sequence have been widely studied in computer vision area for last two decades. Considerable advancement also made in person detection in image such as space time interest point (STIP) by I. Laptev (2005), Histogram oriented gradient (HOG) by N. Dalal and B. Triggs (2005), Bag of words (BOW) by V. Delaitre, I. Laptev, and J. Sivic (2010), Dense trajectory based approach by Heng Wang, Alexander Klaser, Cordelia Schmid, Cheng-Lin Liu (2011), HMM based approach. Most recently CNN based approach by C. Feichtenhofer, A. Pinz, A. Zisserman(2016), G. Gkioxari, R. Girshick, and J. Malik (2015), H. Mahmudul, and Amit K. Roy-Chowdhury (2015), Karen Simonyan, Andrea Vedaldi ,Andrew Zisserman (2013) gained huge popularity due to the outstanding performance in PASCAL VOC and ImageNet challenge by A. Berg, J. Deng, and L. Fei-Fei (2010). These two benchmark dataset contributed a lot to improve the of object recognition in image and video file. Additionally, RCNN by G. Gkioxari, R. Girshick, and J. Malik (2015), Ross Girshick (2015), convolutional two-stream network by C. Feichtenhofer, A. Pinz, A. Zisserman(2016), continuous learning framework by H. Mahmudul, and Amit K. Roy-Chowdhury (2015) played important role while in our work we focus on semi-supervise learning with active learning approach.

Each image file with object bounding box information stored in separate xml file has gain popularity in last several years if we consider the object detection competitions such as PASCAL VOC, imageNet, MSCOCO. Bounding box keeps the region of interest information of particular object in an image. In our work we used Fast-RCNN by Ross Girshick (2015) to detect person.
Action classification specifies the relation between object and person orientation in a scene. In our work we focus on human with nearest object to find out the action type in a picture. There are numerous method have been used to recognize action. However, CNN showed promising result by G. Gkioxari, B. Hariharan, R. Girshick, and J. Malik (2014). In our work, we adopted VGG model where transfer learning played the pivotal role in our experiment. By using transfer learning we can re-train pre-existing neural network (e.g. AlexNet, VGGNet) minor modification and then retrain it in our experiment for action classification by Sinno Jialin Pan and Qiang Yang (2010). It is an efficient way to apply deep learning to solve challenging problem.

3. METHOD

We consider two steps approach for action recognition. First of all, we detect person through Fast RCNN by Ross Girshick (2015) algorithm for each class and then later apply our ASSL framework for increasing human action recognition. Our ASSL method is a state of the art method work in two steps, first step is Semi Supervised Learning (SSL) then in second step we consider Active Learning (AL) method which ensures greater data justification with higher accuracy. Here we apply transfer learning approach where transfer knowledge from earlier learned the task.

Figure 1. Classified action in images with noisy background and shows recognition score at the top.

In this work we propose Active semi supervised Learning (ASSL) framework for action recognition from video and single image as shown in figure 1. We anticipated the action specific person detection where each image contains person information with bounding box data with action stored in xml file. We train the image dataset and make our initial model using openly available VGG16 model and modify the last two layer to adopt for human action recognition. If the initial model cannot predict the right action, we employ active learning method to increase the performance.

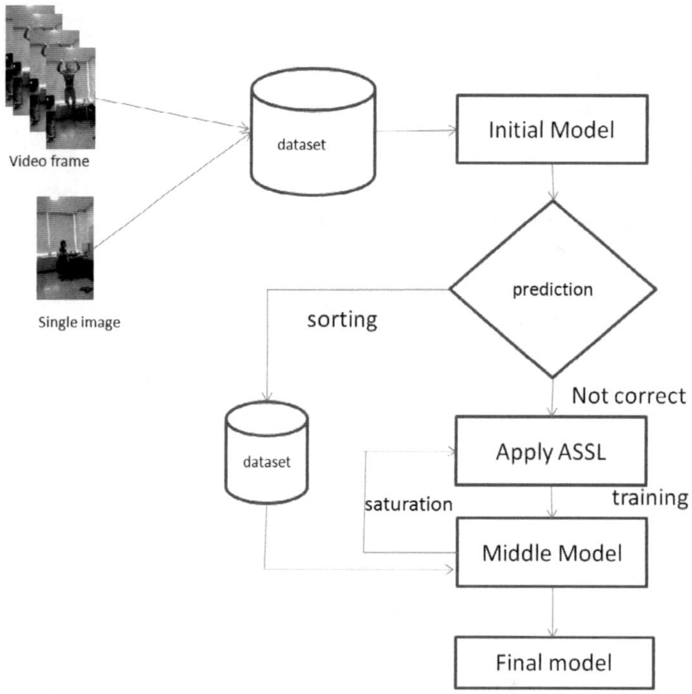

Figure 2. Entire system diagram

We prepare number of training and test image batches, we check the performance against the initial model. For the performance checking we consider threshold value 0.9. If maximum number of image of a batch has wrong prediction as well as less than threshold value we skip that batch and replace with new batch of image and train again. In similar way, we apply incremental learning and prepare the final model which classifies the correct human action.

4. DATASET

The datasets for human action recognition are different from each other due to dissimilar annotation structure. At the same time manual annotation with ground truth is time consuming and tedious work. The pre-trained network was trained on these datasets with image annotation. For action recognition work we use Intelligent Technology Lab dataset. To train the dataset which consist several thousand images where actions are labeled with bounding box information. There are total five different actions such as jumping, phoning, taking photo, using computer and reading. We use the intelligent technology Lab test dataset to evaluate for action classification. The action categorized work presumes knowledge of the ground truth location of the person throughout test time. For the action classification evaluation we use the evaluation criteria defined by the PASCAL VOC by Mark Everingham (2015) action task which computes the AP on the ground truth test boxes.

5. EXPERIMENTS

In this section, we describe the details about our experiments action recognition. We use the CNN architecture by A. Krizhevsky, I. Sutskever, and G. Hinton (2012) which worked extremely well in object classification on ImageNet challenge. We construct on VGG16 network Karen Simonyan, Andrea Vedaldi, Andrew Zisserman (2013) which showed a remarkable improvement from previous state of the art methods.

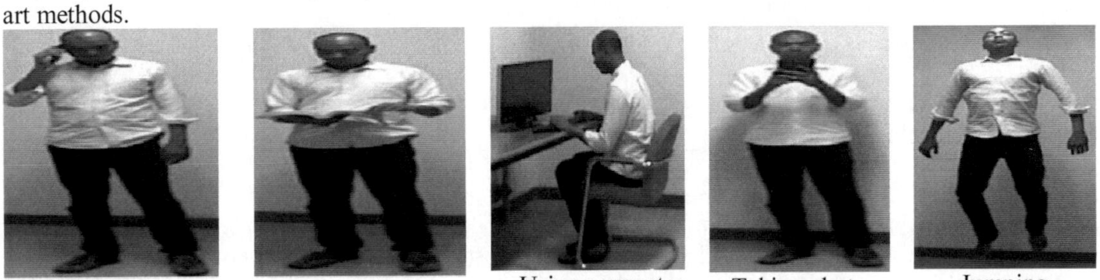

Phoning Reading Using computer Taking photo Jumping

Figure 3. ITLab action dataset where background is simple.

Figure 3. Shows preprocessed action dataset such as phoning, reading, using computer, taking photo and jumping where environment is less noisy. Here pre-processing includes intensity normalization and ROI selection.

5.1 Implementation Details

Our experiment use publicly available VGG net by Karen Simonyan, Andrea Vedaldi and Andrew Zisserman (2013) (http://www.vlfeat.org/matconvnet/pretrained/) that has 5 convolutional layers and 3 fully connected layers. These networks are pre-trained ImageNet for 1000 categories classification task. For the person detection ASSL use the Fast-RCNN by Ross Girshick (2015) that is convolutional neural network based object detector and ASSL frameworks is implemented on the popular deep learning library Caffe. All implementations are on a single server with and CUDA 7.5 with cuDNN 5.1 and a single NVIDIA GeForce GTX 970. We also use Matlab 2015b with matconvnet by Andrea Vedaldi and Karel Lenc (2015) library for both windows 7 and Ubuntu 14.4 operation system in our experiment. We train our initial model for 20k iterations with batch size 8 and learning rate of 0.001. The average single batch of 25 image with 50k iteration batch size 8 using GPU takes nearly 15 minutes.

Table 1. Number of images ITLab Action dataset

	Total	**Average Accuracy**	**Total**	**Average Accuracy**
Training Image	502	94%	621	96%
Test Image	50		50	

Table 2. Number of detected action in the training and test sets in ITLab action dataset.

	Jumping	Phoning	reading	Taking photo	Using computer
Train	100	100	100	100	100
Test	10	10	10	10	10
Correctly recognized image	9	10	10	8	10

We evaluate our experimental result in ITLab dataset. Our propose system, depends on large amount of data for training the model. We compare our results against state of the art methods which show that action recognition method with active semi supervised learning yields superior performance on ITLab dataset.

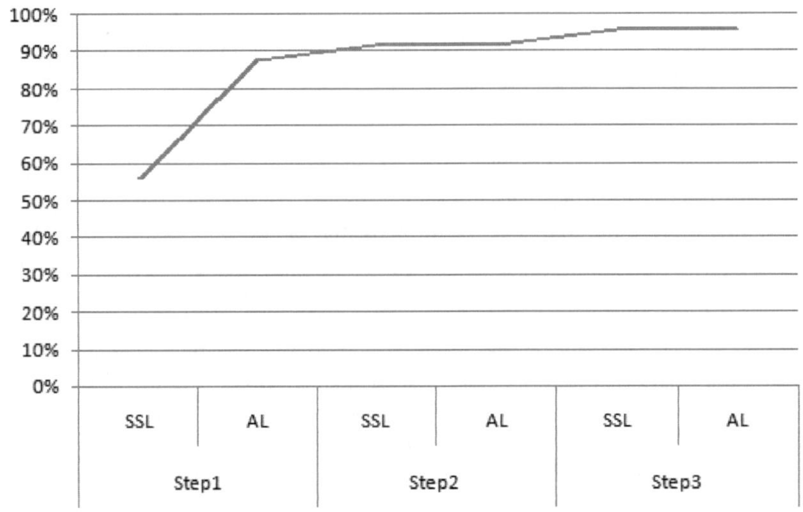

Figure 4. Shows the Semi Supervised Learning (SSL) and Active Learning (AL) steps for Action recognition.

In our experiment, each step consists of two sub-steps AL and SSL. In figure 4. Step one show Semi-Supervised Learning (SSL) get 56% accuracy while Active Learning (AL) obtains 88% correctness. In step two with combined dataset with higher iteration SSL and AL get 92 precision. Finally, in step three both SSL and AL achieve 96% accuracy.

6. CONCLUSION

In this paper we presented a deep learning based framework called ASSL (Active Semi-Supervised Learning) which improves the action recognition performance and reduce the burden of human label effort. Looking at the result, we see that Deep Convolutional Neural Network is suitable for human detection and activity recognition on different ITLab action datasets, but they still require much more tuning with large volume of dataset for better accuracy. In future we would like to compare with other state of the art benchmark dataset with higher performance.

ACKNOWLEDGEMENTS

This work was supported by the ICT R&D program of MSIP/IITP. [2017-0-00543], Development of Precise Positioning Technology for the Enhancement of Pedestrian's Position/Spatial Cognition and Sports Competition Analysis. The GPUs used in this research were generously donated by NVIDIA.

REFERENCES:

C. Feichtenhofer, A. Pinz, A. Zisserman (2016). Convolutional Two-Stream Network Fusion for Video Action Recognition

G. Gkioxari, R. Girshick, and J. Malik (2015). Contextual action recognition with R*CNN. In ICCV.

P. F. Felzenszwalb, R. B. Girshick, D. A. McAllester, and D. Ramanan (2010).Object detection with discriminatively trained part-based models, IEEE Trans. Pattern Anal. Mach. Intell., vol. 32, no. 9, pp. 1627–1645.

H. Mahmudul, and Amit K. Roy-Chowdhury (2015). A Continuous Learning Framework for Activity Recognition Using Deep Hybrid Feature Models. Multimedia, IEEE Transactions on 2015.

A. Krizhevsky, I. Sutskever, and G. Hinton (2012). ImageNet classification with deep convolutional neural networks. In NIPS.

Karen Simonyan, Andrea Vedaldi, Andrew Zisserman(2013). Deep Inside Convolutional Networks: Visualising Image Classification Models and Saliency Maps arXiv: 1312.6034,

Lei Zhang, Zhi Zeng, and Qiang Ji (2011). Probabilistic Image Modeling With an Extended Chain Graph for Human Activity Recognition and Image Segmentation, IEEE transactions on image processing, vol. 20, no. 9

Hanju Li, Yang Yi , Xiaoxing Li , Zixin Guo (2013). Human activity recognition based on HMM by improved PSO and event probability sequence, Journal of Systems Engineering and Electronics, Volume: 24, Issue: 3.

I. Laptev (2005). On Space-Time Interest Points, in International Journal of Computer Vision, vol 64, number 2/3, pp.107-123.

N. Dalal and B. Triggs (2005).Histograms of oriented gradients for human detection, in Proc. IEEE Comput. Soc. Conf. CVPR, Jun. 2005, pp. 886–893.

V. Delaitre, I. Laptev, and J. Sivic (2010). Recognizing human actions in still images: a study of bag-of-features and partbased representations. In BMVC 2010.

Heng Wang, Alexander Klaser, Cordelia Schmid, Cheng-Lin Liu (2011).Action Recognition by Dense Trajectories, Computer Vision and Pattern Recognition (CVPR), 2011 IEEE Conference

Ross Girshick (2015). Fast R-CNN - ICCV 2015

G. Gkioxari, B. Hariharan, R. Girshick, and J. Malik (2014). R-CNNs for pose estimation and action detection. arxiv.1406.5212

LeCun, Y., Bottou, L., Bengio, Y (1998) Gradient-based learning applied to document recognition, Proc. IEEE, 1998, 86, (11), pp. 2278–2324

A. Berg, J. Deng, and L. Fei-Fei (2010). Large scale visual recognition challenge (ILSVRC), 2010. URL http://www.image-net.org/challenges/LSVRC/2010/.

M. Ranjbar, T. Lan, Y. Wang, S.N. Robinovitch, Ze-Nian Li, and G. Mori (2013). Optimizing Nondecomposable Loss Functions in Structured Prediction, IEEE transactions on pattern analysis and machine intelligence, vol. 35

B. Yao and L. Fei-Fei (2010). Modeling mutual context of object and human pose in human-object interaction activities. In CVPR.

Gary Ge, Kiwon Yun ,Dimitris Samaras, Gregory J. Zelinsky (2015). Action Classification in Still Images Using Human Eye Movements, IEEE

Fahad Shahbaz Khan, Jiaolong Xu, Joost van de Weijer, Andrew D. Bagdanov, Rao Muhammad Anwer, and Antonio M. Lopez (2015).Recognizing Actions through Action-Specific Person Detection, IEEE transactions on image processing, vol. 24, no. 11.

S. Ren, K. He, R. Girshick, and J. Sun (2015). Faster R-CNN: Towards Real-Time Object Detection with Region Proposal Networks, arXiv:1506.01497.

Sinno Jialin Pan and Qiang Yang (2010). A Survey on Transfer Learning, IEEE transactions on knowledge and data engineering, vol. 22, no. 10.

Andrea Vedaldi, Karel Lenc (2015). MatConvNet: Convolutional Neural Networks for MATLAB, Proceedings of the 23rd ACM international conference on Multimedia

P. F. Felzenszwalb, R. B. Girshick, D. A. McAllester, and D. Ramanan (2010).Object detection with discriminatively trained part-based models, IEEE Trans. Pattern Anal. Mach. Intell., vol. 32, no. 9, pp. 1627–1645.

Mark Everingham, S. M. Ali Eslami, Luc Van Gool, Christopher K. I. Williams, John Winn, Andrew Zisserman (2015). The PASCAL Visual Object Classes Challenge: A Retrospective, International Journal of Computer Vision

Proc. 1st International Conference on Machine Learning and Data Engineering (iCMLDE2017)
20-22 Nov 2017, Sydney, Australia

Android Based Vehicle Tracking System Using GPS Sensor

Mohammad Sanaullah Chowdhury[1], Nusrat Zahan[2], Israt Binteh Habib[3] and Shamima Akter[4]

Computer Science and Engineering, University of Chittagong, Chittagong, Bangladesh
[1]sana1691@gmail.com, [2]zahannusrat24@gmail.com, [3]isratjahan15.cse@gmail.com,
[4]shamimaakter95@gmail.com

Abstract

Vehicle tracking system is an android based application using GPS sensor which will track and show the nearer vehicles in order to their location. Sometimes it is quite difficult for people to find a vehicle in time of emergency. In this regard vehicle tracking system reduces the suffering of people and also to consume time and energy by creating a contact between the driver and the passenger. Mobile trend and the development of 3G network have changed our lifestyle by making interest in our own smart phone. In such a scenario, mobile application development is one of the most beneficial platforms. Android is one of the largest platforms that run in most smart phone. Vehicle tracking is an innovative android phone based vehicle reservation application which aims to fulfil the demand of user to ease their journey. This paper outlines about vehicle tracking application which helps individuals to hire a vehicle with a phone through sending driver details to passenger. This application helps the user to select a suitable cab and make a phone call to the driver of the cab to pick him up. Java programming language using android operating system for mobile client, PHP as web server, MYSQL as repository, GPS (Global Positioning System) as location provider, etc., benefits are taken in this proposed system.

Keywords: *Vehicle tracking; GPS; Andriod*

1. INTRODUCTION

Vehicle Tracking is based on an android based mobile application and web based management system. The main intention of this system is to render easy use and helpful mobile application to the android users which will make them enable to track a nearest vehicle and reserve sitting from their location comfortably instead of waiting in the road. This system uses smart phone and personal computer in which the first one is used to monitor the vehicle location and the other one act as a tracker. This system uses android mobile phone as mobile terminal because it is more convenient and flexible. Android mobile phone will link to the web server and enter the information. Thus, user can retrieve the information for further action. GPS is used to provide a very accurate location, time, and date. The satellite will transmit the information to the GPS so that GPS can receive the information. By measuring the distance from satellite, it allows the data to identify the location [Kushwaha et al (2011)]. Two types of user can use this application. They are driver and passenger. Drivers have to register the information (name, phone, cab no, email and password). Then driver sign in the information (email and password) that passed to the server. Server checks whether the driver is registered or not. If registered, then next activity will work, otherwise error message will show. Passengers have to register this information (name, phone number, address, email and password). Then email and password passed to the server. Server check whether the passenger is registered or not. If registered, then the passenger can track nearer most vehicle location. Finally, the nearer most driver list is shown and passengers can call the nearest driver.
The users (passengers and drivers) will be much benefited from this vehicle tracking software. This

application will help the passenger (registered) to find the near most vehicles at a short time. It also helps the drivers to find passengers in a short period of times. User friendly interfaces are developed to easy use the software. All information is stored in an organized way easily. This software uses the best searching and finding techniques to enhance the performances of vehicle tracking.

The remaining part of the paper is structured as follows: Section two presents the related works. Section three provides an overview of the related technology. Section four and five presents the architecture, design and implementation of the proposed system. Experimental results and discussion are presented in section six. A conclusion is included to summaries the contribution of the research.

2. LITERATURE REVIEW

Uber is a location-based app that provides hiring an on-demand private driver. Now it operates in 643+ cities in 77 countries. Uber is convenient, inexpensive and safe taxi service for riders. It makes hiring a private driver to pick up passengers and take them to their destination with the tap of a button on any smart phone device. Uber provides drivers an exceptional pay, allows them to be their own boss and even receive tips. After each ride passengers get to rate their driver and overall experience and drivers are reliant on good overall ratings (otherwise they will get the boot). Likewise after each ride drivers get to rate the passenger. So if a rider is troublesome or belligerent, other drivers can get a heads up. Uber is extremely safe for "taxi" service and it has been giving their driver and passenger overall user experience and personal security protocols.

Real time vehicle monitoring system using LCD screens has been proposed for the Mumbai City in India. The main objective of the GPS based vehicle monitoring system is to help the passengers of Mumbai City by giving proper instructions whether to wait for the vehicle or walk or take alternate transport [Chheda et al (2012)].

In SMS Based Vehicle Tracking System, SMS is used over the GSM networks to transfer the coordinates of vehicle location. The vehicle with GPS receiver calculates the latitude and longitude of the vehicle coordinates. This information is directed to the central server using SMS over the GSM networks and is stored in the database. The information can be retrieved by sending the vehicle number and the route number. The arrival time of the vehicle is sent to the user over SMS [Maruthi and Jayakumari (2014)].

Web based application system has been proposed for the Baghdad City in Iraq. This system is developed using the central server system, vehicle-mounted tracking devices and the web based application. Web based application queries about the vehicles real time location and tracks the route on the map through the web application with the embedded Google map and interacts with the database server for the track details. Using the internet access web based application also permits users with diverse operating system platforms to easily query the real time location of vehicle [Salim and Idrees (2013)].

Mobile Application has been proposed for the Pune City in India. Using Android application bus tracking system uses the inbuilt GPS service provided by the Smartphone. In each bus smart phone is mounted to get its GPS coordinates. To the central server these coordinates are transferred. Through android application users can retrieve information about the bus number, route number and the arrival time of the bus. Google maps are used to track the bus and the user graphically [Sardey et al (2014)].

The usage of public transport instead of private vehicles is encouraged in VT and PIS (Vehicle Tracking and Passenger Information System) by APSRTC. VT and PIS (Vehicle Tracking and Passenger Information System) implement APSRTC (Andhra Pradesh State Road Transport Corporation). This system is developed to overcome the crucial problem of road congestion by making it comfortable, convenient, attractive and introducing value added services.

3. RELATED TECHNOLOGY

Different technologies are used to develop this application. The technologies are GPS, Android, XAMPP, PHP and MYSQL. Familiarization of these technologies is given below:

A. GPS

GPS is a radio navigation system that makes allowance for land, sea, and airborne users to determine their appropriate location, velocity, and 24 hours a day, in all weather circumstances, anywhere in the world. GPS stands for Global Positioning System. It is well-known navigation and positioning "technology"—namely the magnetic compass, the sextant, the chronometer, and radio-based devices—impractical and obsolete [Johnson and Russell (1995)]. The GPS satellites are in orbit at 10,600 miles above the earth. Each day every satellite checks the time and position with a ground station and makes any minor correction. By getting bearings from three of the four satellites on the ground, the GPS receiver which contains a computer triangulates its own position. For most receivers, the result is provided in the form of a geographic position - longitude and latitude within a few meters [Brown et al (1995)]. The position can be shown on the map, if the receiver is equipped with a display screen. The GPS receiver may be able to calculate our speed and direction of travel and give us estimated times of arrival to specified destinations, when we are moving. Geographic Information and map making data can be stored by some specialized GPS receivers [Caron and Francois (2006)].

B. Android

The Android operating system is made-up of a virtual machine that runs on the Linux kernel, plus APIs and built-in applications. The open source code under the Apache License is released by Google. Additionally, Android has a large community where developers write application in a customized version of the Java programming language primarily [Meier and Reto (2012)].

C. XAMPP

XAMPP contains Cross platform(X), Apache(A), MariaDB(M), PHP(P) and Perl(P). In both a full and a standard version, self-contained and multiple instances of XAMPP is offered. Without any access to the internet XAMPP is used as a development tool and web server solution stack package to allow website designers and programmers to test their work on their own computers [Dvorski and Dalibor (2007)].

D. PHP

PHP is used in server-side scripting language which is designed for web development and it is also used as a general-purpose programming language. The web server puts together the results of the interpreted and executed PHP code. With a command-line interface (CLI) PHP code is also executed and is used to implement standalone graphical application [Lengstorf et al (2009)].

E. MYSQL

"My while SQL" is the shortening for structured query language. MYSQL is an open source relational database management system (RDBMS) which is introduce in July 2013. It is the world's second most extensively used RDBMS and most extensively used open source client server model RDBMS. [Meloni and Julie (2012)].

4. ARCHITECTURE OF THE PROPOSED SYSTEM

In the following figure, the proposed system architecture has been showed. At first, driver's registration is needed. Then driver can sign-into his account for update his location and availability in such that passenger can track his location. The information of driver is preserved in the server. Then passenger registration is needed and passenger can sign into his account similarly. Now, he enters a distance in which distance he want to find a vehicle. Those vehicle locations are less than the given distance is shown in the vehicle list. Then passenger can call the nearest vehicle.

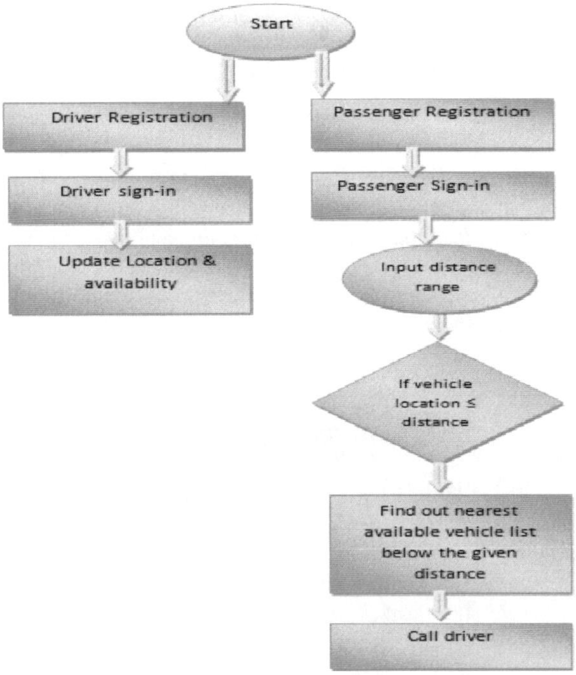

Figure 1. Architecture of the proposed system

In this application, GPS sensor is the root key of the total system procedure. It describes the process of retrieving the GPS location from the user's current location through the android mobile application. For this step, user needs to keep phones GPS on. As soon as, it tracks location at a certain value, it will start its main work on it. The application will retrieve the GPS location of the user's present location.

5. IMPLEMENTATION

We used principle of web languages and tools for the web development and android based languages for making this android application.
The languages are: 1) Java, 2) PHP and 3) MYSQL
The editors are: 1) Android studio and 2) Notepad++.
We have chosen Android studio to implement this application. For android application development, Android studio is the official IDE which based on IntelliJ IDEA. Android studio is one of the top IntelliJ's powerful code editor and developer tools. When building Android application, it offers even more features that enhance productivity, such as:

- A flexible Gradle-based build system.
- Build variants and multiple APK file generation.
- Code templates help to build common application features.
- A rich layout editor with support for drag and drop theme editing.
- Lint tools catch performance, usability, version compatibility and other problems.
- Code shrinking with ProGuard and resource shrinking with Gradle.
- Build in support for Google Cloud Platform, making it easy to integrate Google Cloud Messaging and app Engine.

We have selected notepad++ editor for the server-side scripting language PHP. Our proposed architectural system is server centric and the application server maintains all information in the database. From the server, passengers receive details information of all the drivers on their mobile. Internet works as the medium for transferring user data and service request from mobile to server and the requested information back to the user.

To implement the database, we have selected MYSQL. Here instead of having own database which is stored on the mobile phone we have used central database to reduce the storage overhead on each phone.

6. RESULT

This experiment consists of a single machine which acts as server running windows OS, an android powered smart phone with GPS. Internet connection is initiated between the server and phone. Both the registered users (driver and passenger) can monitor the system.
Register/Login Module: Driver and passenger can register the application by providing the user's information. After the successful registration, they can login by providing email and password as depicted in Figure 2 and 3.

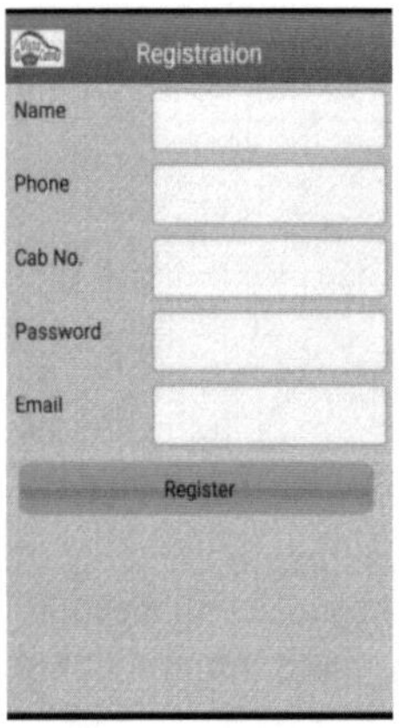

Figure 2. Registration of Driver and passenger

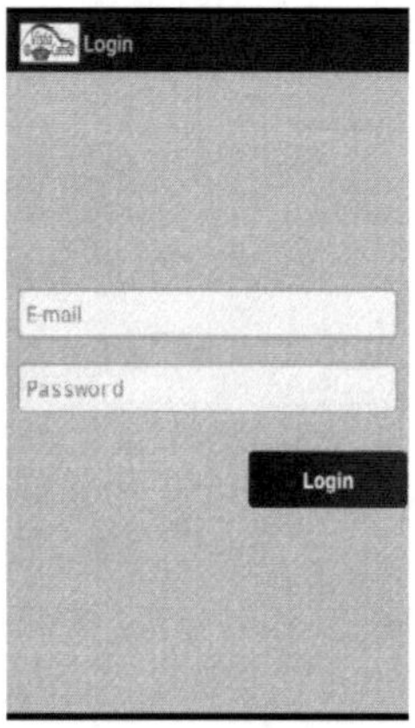

Figure 3. Sign-in of Driver and passenger

Figure 4. Updating location and availability of Driver

Location module: Driver update location and availability as shown in figure 4. Between which distances, the passenger want to get a vehicle, he tab the distance depicted in Figure 5.
Calling Module: The passenger can identify the nearest driver lists between the given distance those are in online that are shown in Fig. 6 and call them to pick him up that shown in Figure 7.

Payment for the trip is calculated as:
Payment = base fare + time rate + distance rate

Price for pick up means base fare. Time includes from start to end of a trip. Distance is calculated as miles or kilometres of route. Example: base= \$3, time rate= \$7, distance rate = \$4. Then total payment will be= (\$3+\$4+\$7) = \$14.

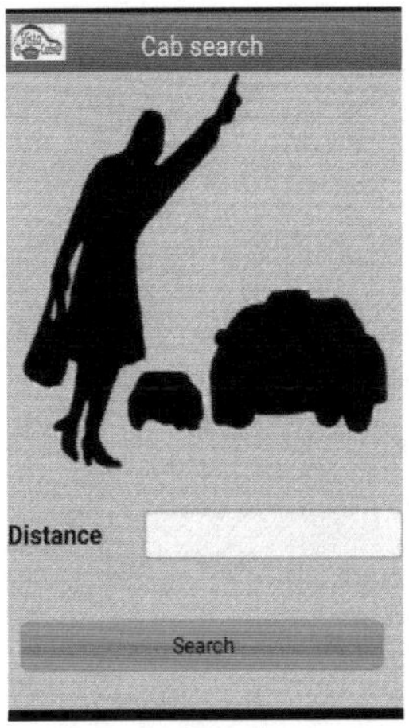

Figure 5. Find out the vehicles between these distances

Figure 6. Driver List

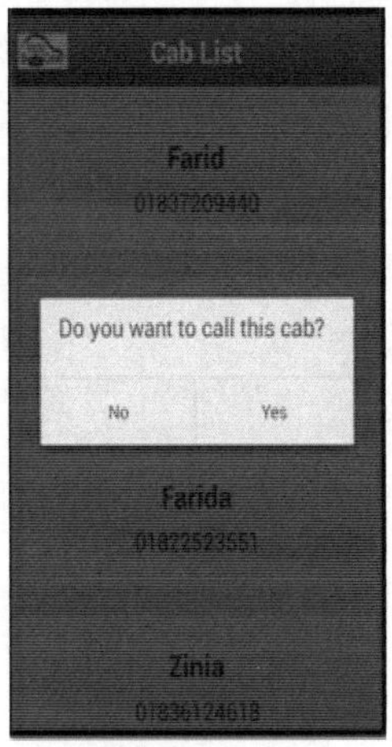

Figure 7. Calling the nearest driver

7. CONCLUSION

The successful creation of fully functioning of Vehicle tracking system provides a feasible solution to manual complex vehicle reservation system. The system allows users to complete online registration rapidly, helps to track the nearest vehicle providing benefits in terms of privacy. This system introduces the detailed process of the Vehicle Tracking System under Android operating system. The prospect for the mobile phone is very bright, particularly for the smart phone with an Android system. Overall this application is user friendly and shows better performance. This application is only usable for one operating system android devices and not usable for iOS devices. We want to implement it with iOS in future. Without internet connection, this system doesn't work. Record keeping is too much analogous in the existing system. We want to extend this research by solving these problems.

REFERENCES

Kushwaha, Amit and Vineet Kushwaha. "Location based services using android mobile operating System." *International Journal of Advances in Engineering and Technology*, 1.1 (2011): 14-20.

Chheda, G., Gajra, N., Chhaya, M., Deshpande, J. and Gharge, S. Real time vehicle monitoring and passenger information system. *International Journal of Soft Computing and Engineering (IJSCE)*, vol. 1, no. 6, pp. 34-58, 2012.

Maruthi, R and Jayakumari, C. "SMS based bus tracking system using open source technologies." *International Journal of Computer Applications*, vol. 86, no. 9, pp. 44-46, 2014.

Salim, A. K. and Idrees, I. M. "Design and implementation of web-based GPSGPRS vehicle tracking system." *International Journal of Science, Engineering and Computer Technology (IJCSET)*, vol. 3, no. 12, pp. 443-448, 2013.

Sardey, Y., Deshmukh, P., Mandlik. P., Shelar ,S. and Nerkar, M. "A mobile application for bus information system and location tracking using client-server technology." *International Journal of Emerging Technology and Advanced Engineering*, vol. 4, no. 4, pp. 86-91, 2014.

Johnson and Russell, K. "Separated GPS sensor and processing system for remote GPS sensing and centralized ground station processing for remote mobile position and velocity determinations." U.S. Patent No. 5,420,592. 30 May 1995.]

Brown, Alison, K. and Sturza, M. A. "GPS tracking system." U.S. Patent No. 5,379,224. 3 Jan.

1995.

Caron and Francois. "GPS/IMU data fusion using multisensor Kalman filtering: introduction of contextual aspects". *Information fusion*, 7.2 (2006): 221-230.

Meier and Reto. Professional Android 4 application development. John Wiley & Sons, 2012.

Dvorski and Dalibor, D. "Installing, configuring, and developing with Xampp." *Skills Canada* (2007).

Lengstorf, Jason, and Hansen, T. B. "PHP for absolute beginners". Vol. 1. Apress, 2009.

Meloni and Julie, C. "Sams teach yourself PHP, MySQL and Apache all in one". Sams Publishing, 2012.

Proc. 1st International Conference on
Machine Learning and Data Engineering (iCMLDE2017)
20-22 Nov 2017, Sydney, Australia

Biometrics and data security: An investigation on biometric templates in computing environment

Roshan Dhakal[1], Rafiqul Islam[2] and Champake Mendis[3]
[1]Student, School of Computing and Mathematics, Charles Sturt University,Wagga Wagga, NSW, Australia
[2]Lecturer, School of Computing and Mathematics, Charles Sturt University,Wagga Wagga< NSW, Australia
[3] Lecturer, Computing in Mathematics, University of Melbourne, Parkville, VIC, Australia

Email: droshan_2005@yahoo.com;mislam@csu.edu.au;cmendis@csu.edu.au

Abstract

Much research on biometric technology and information system focuses on privacy and security such as access control and information security. Few researches infer issues and challenges in biometrics and how they relate to privacy and security within biometric technology. Thus, this research investigates how issues and challenges in biometric are correlated with privacy and security of data and information in biometric technology and organizational information system network. The outcomes would assist organizational management to determine whether implementing biometric technology invades. A survey that has been conducted on fingerprint and facial recognition (aged in between 20-50) found that there is a substantial relationship between biometric user authentication system and privacy and security of user data and information. Additionally, the implementation of biometric technology for user authentication and access control draw the attention and curiosity and this technology were not apparent to be depraved, abusive and fierce. Biometric technology that contains user information is more efficient in identification and verification purpose and does not need to remember like password but once stolen could not be replaced. The outcomes further point out that biometric data and information are irreplaceable and once compromised, then cannot be replaced or changed like passwords. Biometrics technology is apposite and user friendly even though the contentious stratagem of data uniqueness and its cost of stolen. Organization management should therefore carefully understand the need of organization and the feature and risk factor of biometric technology should be considered while implementing it.

***Keywords**:* Biometric technology, privacy and security, identification and verification, user authentication, access control

1. Introduction

Different biometrics technology ultimately differs in its primary technologies, complications and presentations. Every biometric technology has its own benefits and drawbacks and it is very hard to determine such biometric technology which fulfils all the technological needs. Therefore an in depth investigation on analysis of security and privacy issues on each biometric technology and potential solution is vital in organizational information security policy and decision making before and all the way through the implementation of biometric technology. From time to time, possible settings have been worked out and distinct way out is accessible in border security and business purpose. On one hand, the curiosity of biometrics technology has been widespread for its benefit, but on the other hand the possible threats on the user's privacy that arises from the misuse of biometrics information often thwart the implementation of biometric technology in wide ranging. In reality, people are not passionate to hand out biometrics actions with self reliance that biometric data and information cannot be lost and manipulate without user's permission. This is the main reason that people are more concerned with implementing the biometrics technology in reality and also the cost of recovering the lost or stolen data and information and compensation for the breaching the privacy of the user is another factor. Currently, more research has been committed to building the method for securing the biometric data. Like this, user authentication in biometric technology can be worked out and fulfilling the user's concern in relation to security and privacy of biometric data. This kind of procedure generally allows modification of entered biometric characteristics into secure identifiers and turning out to be impracticable to retrieve the original biometric characteristics. Numerous suggestions have been put together linking encryption technology and biometric technology to improve the certainty while biometric data are collected and stored for verification purpose (Cimato, Gamassi, Piuri, Sassi, & Scotti).

The main purpose of this paper is to recognize the organizational needs in relation to access control applications and also assisting in finding the proper biometric technology which satisfied the organizational requirements. Alternatively, this paper also assists in discovering the further and other security control for efficient information security, sovereignty that arises by assessing the risk factor from methodological design and procedural issues (Cimato, Gamassi, Piuri, Sassi, & Scotti). Many literatures have been detailed, documented and focused on the benefits of biometric technology implementations such as its user friendliness, easy to use, security, legal and ethical aspects, etc. An analysis on technical factors of biometric technology has been testified in smaller number and furthermore which are presented is based on technological factors and purposeful matter and still not inclusive (Cimato, Gamassi, Piuri, Sassi, & Scotti). This paper presents the well throughout up to date analysis on privacy and security issues of biometric technology, which covers technological factors that are connected to biometric functioning influenced by various mechanisms that develop the system.

This paper further exposes that challenges on the strength of the system difference in the real time working environment and the security vulnerabilities of spoofing and repeated attacks remain the main problem on top of the biometric technology application. This paper also highlights on the necessity for consistency biometric system analysis, reporting and guarantee of biometric device compatibility. The research paper is structured as follows: Section 2 represents Biometrics performance methods; Section 3 represents Experimental study and measurement; Section 4 represents Results and Analysis; Section 5 represents Discussion and future direction; Section 6 represents the conclusion.

2. Biometrics performance methods

The primary purpose of biometrics technologies are identification and authentication and both needs the subsistence of user's reference data which will be equated to reference template or odd data. Before discussing the identification and authentication phase, enrolment process needs to be defined. Enrollment phase in biometric system inputs the user's biometric data and characteristics into the database of biometric system. In enrollment stage, the user places his finger, face or hand or retina over a biometric device or reader, which scans the user's biometric features and creates a digital symbol or icon. The biometric system normally does a quality assessment to make sure that consecutive stage could develop the acquired sample consistency. For easy matching, a feature which is named as extractor processes the entered sample to create a compress other than meaningful image known as a template. Biometric system could store these templates on the smart card that has been assigned to user or register these templates into a central database which relies upon the biometric applications (Sharifah Mumtazah Syed Ahmad, 2012).

Biometric identification means the capability of the computer information system to exclusively differentiate the user from stored biometric templates in the database. For example, the government of any country implements the national biometric identification system which admits the residents to confirm his/her identity without presenting any document and believing that residents are previously recorded in the national biometric identification system. In other hand biometric data will be compared with all other templates that are stored in a database on a national biometric identification system and when a successful match is found with the corresponding resident's identity then data would be obtained from the database. This process is also called as "one to many "match and usually employed by police and forensic department to identify the criminals (Cavoukian & Stoianov, 2007). Whereas biometric verification means "one to one" match where real time biometric data presented by the user is matched with the stored biometric templates that has been previously provided by the user and the match is established. Here biometric devices could match the real time biometric data and there is not necessary to search the biometric templates in the database even though database could be used. This kind of comparing the real time biometric data to the recorded biometric template is all the essential requirements to authenticate the user is correct user (Cavoukian & Stoianov, 2007). Verification need not insist identification all the time when the appropriate user uses the facility.

Biometric performance is based on the two results known as "false accept" and "false reject" and these are two common mistakes that happen throughout the identification and verification process. The degree which measure these ensue is called as "False Accept Rate" and "False Reject Rate" which is used to calculate the level of correctness and consistency of biometric systems. While measuring the user's biometric data, the biometric data is matched to the enrolled biometric templates and if the biometric system fails to match, for instance, user data have been stored in biometric template and still biometric system fails to recognize then it is called "False Reject Rate". For a particular biometric system, the False Reject Rate would be the similar process for verification and identification. External factors such as the user's wet or dirty hand, poor lighting for Irish recognization, etc. could affect the False Reject Rate (Navaz, Sri, & Mazumder, 2013). On the other hand, when user's characteristic is satisfactorily identical to the stored template and the result match is false stated then it is known as "False Accept Rate". The False Accept Rate of the given biometric device precisely manifest to the consistency and the influence of the equipment. To receive the low False Accept Rate user's data must be totally distinctive and the algorithm employed should summarize record and seek and return user's distinctiveness in an efficient manner (Navaz, Sri, & Mazumder, 2013).

Additionally, the term "collectability" also needs to be addressed which is another significant feature of biometric systems. While all biometric technologies rely on the input devices that focuses on active command, in some cases where the devices itself fails to acquire or capture the required entered samples in satisfactory circumstances. In that case, it might require the collectability necessity which in return delays the recognition and this process is known as "Failure To Acquire" or "Failure To Capture". For instance, application of residents for biometric smart card needs high output, demand for low Failure To Acquire. Beside these facts, a real biometrics technology should comply with the comprehensive prerequisite which basically means that it is should be user friendly and practical for all users (Cimato, Gamassi, Piuri, Sassi, & Scotti).

3. Performance based on literature review

The literature review is involved in the study of previous research and identifying and comparing the difference in previous and current studies. According to Irfan Iqbal (McConnell), most of the researcher believes that biometric technologies provide a high level of security and privacy, but researchers' crusade with the truth that it is hard to show the return on investment of biometrics technology (McConnell). Is the biometrics technology fulfilling the organization's needs? How much will be the ongoing cost for maintenances? Biometrics relate to the personal data and information, so there is concern in relation to privacy and security of this data and information from a legal point of view (McConnell). After collecting the biometric data and information from the users, these data and information as well as biometric devices needs to be protected. The only solution for this is to encrypt the biometric data and store in central database rather than local workstation. Users trust and awareness in relation to biometrics technology could affect the successful execution and competence of the biometric technology (McConnell). Some of the other issues could arise in the working environment that affect on the performance of biometrics technology. For example, wet and dirty hand could affect in reading the finger scan and hand geometry, poor lighting could affect in retina scan and facial recognition, extra makeup could affect in detecting faces recognition, any cuts or wounds in hand or palm could affect in detecting finger scan and hand geometry (Iqbal, 2017). In order to address such issues, alternative way for access control should be provided such as pin number or password. Training and awareness should be provided to users about handling the biometrics technology.

Even though, the extensive research in relation to privacy and security issues in biometric has been done and yet additional research needs to be carried out. As discussed earlier that security and privacy are directly related to the single user and also the effectiveness of biometric system implemented for user authentication. Further, in depth analysis points out that there is some section on biometric system that needs urgent concern. Fingerprint was believed to be one of the best and most secure biometric tools, but due to increase in technology and digital biometric devices finger print scanning could be fooled. And there are some other websites which provide the full step by step information which can generate the duplicate finger print. Facial recognition was also believed to be a secure method of biometric tools, but it also could be deceived with a mask at non attendance circumstances. Now a day, cosmetic make up and plastic surgery are other methods to deceive the facial recognition. Additionally, faces transforms eventually, the recorded templates has an inadequate time period that influences biometric system consistency. Moreover, poor lighting condition and crash in the whole system are other issues related to privacy and security (Zimmerman, 2002). Voice reorganization could be tricked by manipulating the recorded voice with the voice recognition software and matching them with an original voice. Irish cameras are very expensive that require high technology and also are not suitable in every environment. The primary issue is that if other technology provides high speed data and template matching then why there is a necessity if Irish recognization (McConnell)? Biometric data and information once compromised would not be able to change like passwords. How users are convinced that the compromised data would not affect their privacy? None of the biometrics is 100% secure except DNA matching which is highly costly and is not ethical to implement this for user identification and authentication. Biometric system does not provide any security of the user so why an organization should implement it for access control? The cost of employing biometric should and its ongoing cost of maintenance of biometric system and cost for protecting the template data are other issues. Spoofing attacks, the man in the middle attacks, replay and substitution attacks, Trojan horse attacks and masquerade attacks are some of the common attacks in relation to security and privacy of biometric system.

After reviewing literature on implementing the biometrics technology, some of the gaps have been identified in research in relation to factors affecting whether to employ biometrics technology or not. Study and investigation in this field help organization management to know organizations could be beneficial by implementing the biometrics technology. Moreover, it will also helps information technology professionals, information security managers and analysts to decide which area of biometrics technology they are related to and also know which biometrics solution suits best for their organizations. Some of security organizations which work in information security could also gain advantage from this paper for proposing information security solutions (McConnell).

4. Performance based on secured biometrics templates

Biometric templates are generally altered prior to store during the enrollment stage so that biometric user authentication process could be carried out accurately, but illicit access to the stored templates allows the foe with the fewest amount of real data on the biometrics of the confronted user. The usual method for securing the templates is to reproduce the method that is used in a password based authentication method. In this method, passwords are stored in hashing form and hash function never provides any information and in the event of system crash password is not leaked (Zimmerman, 2002). Another criterion for every created biometric template is known as unlikable characteristics which underlines its originality. This is crucial to enhance the security features mostly since the similar biometric characteristics could be exercised to recognize the user in several applications (Cimato, Gamassi, Piuri, Sassi, & Scotti). For instance, suppose biometric application such as a fingerprint is used to gain access to secure building and through the biometric template of smart card and same fingerprint application acts as access control to the central database. So if the hacker finds that smart card he might manage to get access to secure building, but could not gain access to central database because the fake smart card hacker generated would not have administrative rights into account. Renewing the biometric templates, which allows replicating another template while the current templates are leaked. Renewable features basically need the reversible possessions that permit biometric template to reissue in the same way that compromised templates are replicated. This feature is also implemented while cancelling and deactivating the compromised template without in disrupting the new substitution. In biometric encryption technology, a digital key is randomly created during the enrollment stage, even though the user does not know about it. When the biometric sample is captured, then biometric encryption algorithm securely combines the key to biometric sample to generate a biometric encrypted template. Therefore the key is considered as an encrypted key with the biometrics. While verifying, the users entered the biometric sample, then it is matched with the encrypted biometric template. At that same time the key acts as decryption key (Cavoukian & Stoianov, 2007). Implementing the multiple security policy or defence in depth technology is another solution for effective biometric technology. For instance, to gain access to highly classified information such as the bank central locker there should be multiple biometrics technology for identification and verification so that there would be no compromise in privacy and security. In this case implementing the multiple biometric techniques such as fingerprint, heart beat and facial recognition personally at the same time would be effective because if hacker creates the fake fingerprint and even the facial mask then it would not be possible to get biometric information on the heartbeat.

5. Experimental study and measurement

In this paper, the performance of biometrics system is measured in terms of FAR and FRR for authentication accuracy and protection of biometrics templates. The experiment is measured by how the experiment is carried out and how the performances of biometric templates are measured.

5.1. Conducting experimental outline

In order to investigate the research question and sub questions, an appropriate framework for research is needed. There are so many methods and models that could be implemented in this research, such as qualitative, quantitative, mixed, etc., but quantitative research methodology has been implemented in this research. The main reason to choose quantitative is that the survey will be used as a tool for data collection and the result of this research will be compared with the previous for reliability and validity. There will be questions related to encryption, detection and security and privacy which will be easier to get results in the numbers such as in the scale of 1-10 where 1 is low and 10 is high and collecting data through multiple choice questions where answers are provided in option rather than participants saying that I have no idea about the topic you are talking about. The research initiates from stating the problems along with the objective of the research. Literature review has been conducted to identify if there is any gap in relation to this research and how this research fulfil those gaps. The survey will implement as a tool for data collection from the organization and the survey questions will be related to organization, attitude and opinion on biometrics security and privacy and also to ensure whether organization are interested in long term investment in biometrics technology. A survey would be the worthy research methodology for solving our research questions and sub questions. In questionnaires, some of the factors such as cost, user friendly, easy to use, accuracy and reliability and other desired security and privacy level could be measured which other research may or may. Online survey would be the most effective for this research and the security and privacy of participants would be kept in mind because in the interview there would be face to face interaction within interviewer and participants. While collecting data there are few things that need to keep in mind such as the level of user understanding of biometrics technology, security incidents occurred within biometrics systems and there should not be any offensive questions such as have you ever created security incidents in organization during work?

6. Performance calculation of biometrics templates

Biometrics templates are calculated based on the False Accept Rate (FAR), False Reject Rate (FRR), Failure to Enroll (FTE) and Failure to Capture (FTC) performance. FAR calculates the rate of users in percentage that are successfully admitted as real users and FRR calculates the user's percentage rate who are admitted as real users but rejected during the authentication process. Both FAR and FRR could normally be substituting adjacent to themselves by modifying certain factors. Although sometimes, both FAR and FRR are identical commonly known as Equal Error rate (EER). The biometrics system could be more precise if there is low EER. To get the low false rate biometrics system could imply certain factors such as biometrics devices are programmed with the high acceptance limit. For instance, in any biometric device three attempts are allowed to minimize FRR which means users are falsely rejected if user fails all three attempts for authentication (Prabhakar, Pankanti, & Jain, 2003).

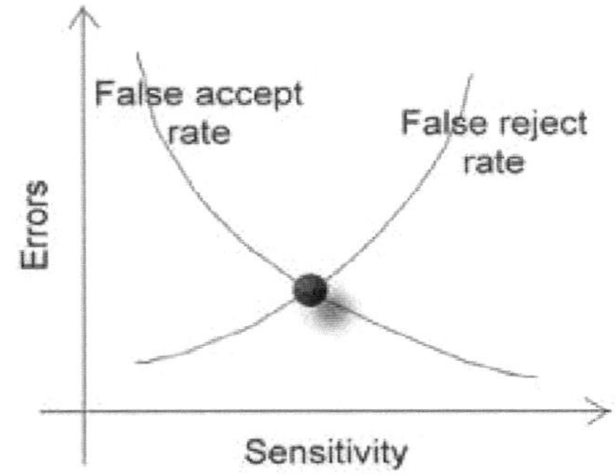

Figure 1: Calculating Biometrics templates performance (Prabhakar, Pankanti, & Jain, 2003)

7. Results analysis

To explore the performance on verification of biometric system, it should be viewed on how biometrics systems responses to enormous biometric features inquisition from accepted users and unaccepted users. As a result of physical instability and limitations in computation, the outcomes of such research are indistinct and could be only forecasted to specific limit. To resolve the error rates which is false accept rate and false reject rate, the users judgement of yes or no are not accepted rather than principle extent of equality in between inquiry templates and stored reference templates in database(Biometrics performance determination for specialists, 2017). For large number of samples, the equality ratings which is also known as score values are gathered for both accepted and unaccepted users and frequency of prevalence is calculated for each score values. As soon as it becomes constant with the entire inquiry templates, an accepted and unaccepted user's statistical outcome creates estimation to the probability distribution function which gives the clear view of quantified approximation of particular equality ratings number of probability (n) of occurring for accepted users (pB(n)) and unaccepted users (pN(n)).

$$pB(n) = \frac{\text{total number of accepted users with same ratings}}{\text{total number of accepted users}}$$

$$pN(n) = \frac{\text{total number of unaccepted users with same ratings}}{\text{total number of unaccepted users}}$$

There will be more exact estimation when there are high volumes of accepted and unaccepted users. Sometimes there is no traverse or overlapping in the statistical distribution curves which means high equality ratings inquiry templates for accepted users and low equality inquiry templates ratings for unaccepted users (Biometrics performance determination for specialists, 2017). In such condition, decision threshold is used to identify the division in between accepted and unaccepted users but in realistic both distribution curves overlap because of large number of users.

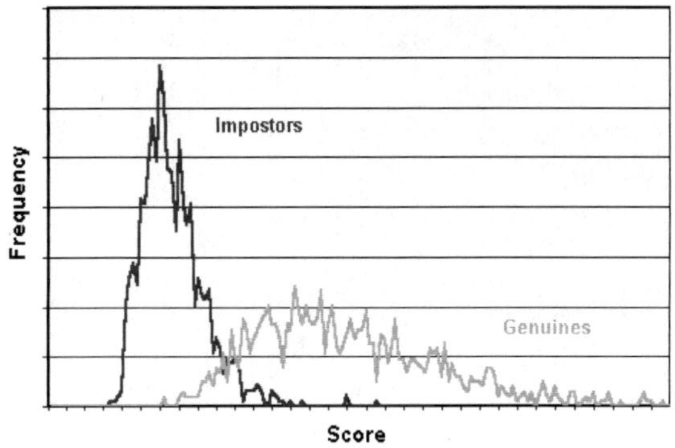

Figure 2: Frequency diagram of genuine and impostors

The error scale values of FAR is described as the probability of unaccepted users is accepted as authorized users and FRR is described as accepted user is unaccepted as unauthorized user. The distribution curves depend upon the amendable decision thresholds for comparison of scanned biometrics templates with the stored reference templates from database (Biometrics performance determination for specialists, 2017). The pursued source is directed below the approximation which is the equality rating value could be any number in between 0 and K and for easy favour the starting probability value of K would be 0.

When the common probability distribution function p is known for distinct alike values n then probability pM(th) of the scanned biometric templates with same rating n comes under threshold th9"misses)which is

$$PM(0) = 0$$

$$PM(^{th}) = \sum_{n=0}^{th-1} p(n) \qquad th=1, 2... K$$

The total of accurate alike and dislikes should be identical to the total number of events. Therefore the probability PH (th) which has the same rating of the scanned biometrics templates attain or go beyond the thresholds th ("hits") is

$$PH(^{th}) = 1 - PM(^{th}) = Kp(n) \qquad th=1, 2, ..., K$$

$$\sum_{n=0}$$

For the False Match Rate FMR(th), the probability approximation which have same couple of separate templates neither could not attain the definite threshold value nor go beyond the definite threshold value th.

$$FMR\ (^{th}) \sim PH\ (^{th}) = 1 - \sum_{n=0}^{th-1} pN\ (n) \qquad th=1, 2, ..., K$$

And similarly for False Non Match Rate FNMR (th):

$$FNMR\ (^{th}) \sim PM\ (^{th}) = 1 - \sum_{n=0}^{th-1} pB\ (n) \qquad th=1, 2, ..., K$$

Where pN =probability frequency function for unaccepted users

And pB=probability frequency function for accepted users

The symbol $\sim$ denotes that the normal value of measures=d failure rates False Match Rate and False Non Match rate are similar to the probabilities PH and PM respectively and the limit values are as follows:

FMR (0) =1 FMR (K) =0

FNMR (0) =0 FNMR (K) =1

In order to measure the FAR and FRR, the threshold own Quality Rejection Rate which is same as Failure To Accept should be measured. While given false acceptance is allocated to false match then it will be achieved:

$$FAR\ (^{th}) = (1-QRR)\ FMR\ (^{th})$$

$$FRR\ (^{th}) = QRR + (1-QRR)\ FNMR\ (^{th})$$

And for circumference value it will be achieved:

FAR (0) =1-QRR FAR (K) =0

FRR (0) =QRR FRR (K) =1

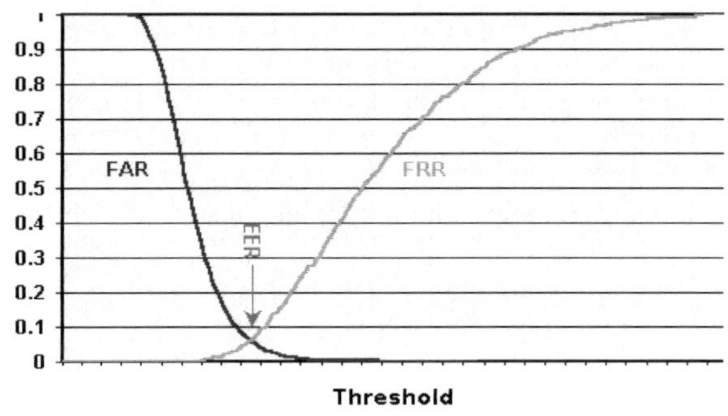

Figure 3: FAR and FRR graphical analysis

8. Discussion and future direction

Whenever there is an inside or outside attack, biometrics technology does not provide protection in authentication system by users. Here are some of the findings that organizations could implement in order to safeguard biometric authentication system. A database that stores biometrics data and information need to be encrypted with encrypted key. High priority should be given to audit logs and full attention should be given to detect if there is something wrong user does or happens which could compromise privacy and security. Standard logs should be put into effect which can identify the user who is not complying with the policy and procedure that may compromise the biometrics system. Biometrics encrypted data and information should be located in the central database rather than local workplace. This helps encrypt data and information less chance of interference and mess about (Zimmerman, 2002). And finally the most important is to educate and aware the user about biometrics technology such as how it works, what will happen when it gets breaches and what would role of user when there is intruding in biometrics technology. This paper also finds that extra biometrics technology should be implemented where possible. The analysis of security and privacy in relation to biometrics implementation find out that there are critical areas which needs more research and needs more focus beyond the technological limitation. Further research needs to be carried out to find out the factors that are responsible for the failure to capture during the enrollment process. There is no proper framework for measuring the success factor of biometric templates rather than false accept rate and false reject rate, therefore additional research need to be carried out. More research should be carried out in anti spoofing procedure which aspiring to create the smart biometrics components: identification and verification that differentiate the genuine and forged sample. More research should be carried out in live monitoring through security cameras and identifying and detecting the circumstances of biometrics presented from simulated devices.

9. Conclusions

The biometric system saves time during user identification and verification; the user does not have to remember the password and they do not have to hassle for changing password. Organization management should understand the need of organization, calculate the risk factor and must understand which kind of biometrics would satisfy the organizational needs prior to deploy biometrics system. There is no risk in implementing the biometric system, but primary concern is how these installed biometric systems should be protected so that it protects our privacy. The organization should always look for another option while in case of emergency so password policy also should not be forgotten as plan B.

In conclusion, this paper addressed some of the issues that arise in biometric privacy and some potential solution to address those issues. The purpose of this was to introduce the biometric system in real time environment for identification and verification process. The lawful execution of the biometric measures assists some of the families that have been identified and reunited with their parents and children through biometric DNA matching who were missing since the Second World War. Another example could be the death declaration of Saddam Husain and Bin Laden was identified by biometrics measures DNA because US military have got samples of those high profiles. Biometrics system would gain more and more interest and it will be the only most efficient measures for identification and verification.

References

Ann Cavoukian, A. S. (2007, 3). Biometric Encryption: A Positive-Sum Technology that Achieves Strong Authentication, Security AND Privacy. Accessed via www.ipc.on.ca/wp-content/uploads/Resources/bio-encryp.pdf on 27 August 2017

Zimmerman, M. (2002). Biometrics and User Authentication. SANS Institute: Accessed via www.sans.org/reading-room/whitepapers/authentication/biometrics-user-authentication-122 on 28 August 2017

Biometrics Performance Determination (for Specialists). (2017). Biometrica: Accessed via www.bio-metrica.com/biometrics-performance-determination 14 September 2017

Cimato, S., Gamassi, M., Piuri, V., Sassi, R., & Scotti, F. (n.d.). Privacy in Biometrics. Accessed via http://citeseerx.ist.psu.edu/viewdoc/download?doi=10.1.1.464.649&rep=rep1&type=pdf on 26 August 2017

Iqbal, I. (2017). A Proposal for Factors Influencing Biometrics Technology Security. International Journal of Innovative Research in Computer and Communication Engineering , 3651-3657.

McConnell, J. M. (n.d.). Understanding biometrics: How to choose the right biometric technology for your organisation. Planet biometrics. Accessed via www.planetbiometrics.com/creo_files/upload/article-files/how_to_choose_the_right_biometric.pdf on 28 August 2017

Navaz, A. S., Sri, T. D., & Mazumder, P. (2013, 3). Face recognition using principal component analysis and neural networks. Research gate. Accessed via www.researchgate.net/publication/235950165_Face_Recognition_Using_Principal_Component_Analysis_And_Neural_Networks 13 September 2017

Prabhakar, S., Pankanti, S., & Jain, A. K. (2003, 3). Biometric Recognition: Security and Privacy Concerns. Accessed via https://www.researchgate.net/publication/3437477_Biometric_Recognition_Security_and_Privacy_Concerns 28 August 2017

Sharifah Mumtazah Syed Ahmad, B. M. (2012). Technical issues and challenges of biometric applications as access control tools of information security. International Journal of Innovative Computing, Information and Control , 7983–7999.

Stelvio Cimato, M. G. (n.d.). Privacy in Biometrics. Accessed via http://citeseerx.ist.psu.edu/viewdoc/download?doi=10.1.1.464.649&rep=rep1&type=pdf on 26 August 2017

Zimmerman, M. (2002). Biometrics and User Authentication. Accessed via www.sans.org/reading-room/whitepapers/authentication/biometrics-user-authentication-122 on 29 August 2017

Proc. 1st International Conference on Machine Learning and Data Engineering (iCMLDE2017) 20-22 Nov 2017, Sydney, Australia

Information security training and awareness program: An Investigation

Roshan Dhakal[1], Rafiqul Islam[2] and Champake Mendis[3]
[1]Student, School of Computing and Mathematics, Charles Sturt University,Wagga Wagga, NSW, Australia
[2]Lecturer, School of Computing and Mathematics, Charles Sturt University,Wagga Wagga< NSW, Australia
[3] Lecturer, Computing in Mathematics, University of Melbourne, Parkville, VIC, Australia

Email: droshan_2005@yahoo.com;mislam@csu.edu.au;cmendis@csu.edu.au

Abstract

Information security training and awareness program is a bottom line in preventing the information security incidents in an organization. The success factor of this program would be unknown unless its efficiency is measured. Prior to implementing the information security training and awareness program, organization management should identify the organizational needs, metrics to measure the efficiency of the program and update the activities in the program. Information security training and awareness program teaches the employees how their knowledge, attitude and behaviour affect organization's overall performance. This paper is not proposing any specific information security products, but giving practical suggestion to organizational management and users regarding information security. This paper also proposes that it's everyone's responsibility to secure organization's information system. This paper measures the success factor of the information security training and awareness program and ensures organization management that investment in such program is valuable in long-term. This paper discussed on research methodology, outlining the experiment and statistical analysis and measuring the effectiveness of information security training and awareness program. This paper further analyse on some other factors that contributes in making training and awareness program effective and also provide suggestion for future research.

Keywords: training and awareness program, metrics, knowledge, attitude and behaviour

1. Introduction

Today information technologies are all reliant on users and information security is all about the user's behaviour rather than technical measures. Users planned actions or accidental actions can tempt undesirable results which information security experts to prevent. Regardless of the promotion about the requirement of information security products from stakeholders, there are numerous sensitive information security, a performance that is not and could not be computerized. Information technologies are designed to permit an information security in a long run, which depends upon the successful employment and undertaking those technologies by qualified users. This signifies that organizations are heavily dependent upon the users in order to carry out the safe environment. Employees are still believed to be the weakest link in an organization and are the biggest threats to organizations. Organizations should implement enough policy to ensure that their employees are getting proper knowledge on organizational information security policy and procedures (Stephanou & Dagada). The aim of educating and training users is to make sure that the user implements essential policy and procedures and also user are not abused or users do not mishandle the organization's information security policy.

According to Sans (Russell, 2002), the aim of the information security training and awareness program is to teach users about their roles and responsibilities in securing confidentiality, integrity and availability of an organization's information and properties. Therefore, information security is not just about the responsibility of information technology or any information security department, but it is the responsibility of all users. So it is necessary to know by every user how to secure the organization's information system and also what happens when users fail to do so?

According to ABC news (Edwards, 2015), more than three-quarters of a million Australian populations (3.1% of the total Australian population among the age group in between 25-34 years) have been a victim of identity theft. And according to the Australian Bureau of Statistics (Identity Theft, 2016), one-fifth of identity theft victims have been encountered same incidents again and again in previous years (Braue, 2014). The common way that identity theft happens in Australian is through phone calls, messages, email and internet, door to door salesperson, etc. Among these victims around half percentages does not know how their personal details have been collected.

It is now clear that information security training and awareness acts, as the first line of defence for any organization which brings down the organization's information security threats and risks into the acceptable level. And today every organization has its own organization information security policy and procedures and training and awareness program. But how does the organization know that the training and awareness program they are implementing is achieving the goals unless measures the effectiveness of training and awareness program (Stephanou &Dagada). The main purpose of this paper is how the efficiency of information security training and awareness program is measured in the real computing environment.

Furthermore, this paper tries to find the success factor of the training and awareness program. This paper also assists organizational management to decide whether the training and awareness program that has been implemented is the right or needs to reassess (Measuring Outcomes). After gathering the information and knowing the effective methods, an organization can decide which area needs more attention and focus so that ongoing funding would be supplied. This paper is structured as follows: Section 2 represents Background; Section 3 represents Motivation; Section 4 represents Related Work; Section 5 represents Methodology; Section 6 represents Finding and Discussion; Section 6 represents Conclusion.

2. Background

Measuring the user behaviour is anonymous that cannot be considered entirely. User's thinking level, understanding level and the aptitude of transforming and confronting is not same. Therefore, to combine users with all these attributes, there is only one formal document which is known as information security policy and procedures (Brodie, 2008). There could be confusion that information security policy and procedures regarding job performance (which means that information security policy and procedures do not teach others how to do the job, but it does teach how users can perform their job safely and securely) (Giovanni, 2015). Moreover, information security policy and procedures also teach users about the result to arise from doing the job inappropriately, how to report any incidents and users' rights in an organizational information system (Maqousi, Balikhina, & Mackay,

2013). In information security education, training and awareness program, there are three attributes: information security education which teaches the users on organizational information security policy and procedures, information security training trains users how doing the job safely and securely and information security awareness which is designed to change the performance of the users (Stephanou & Dagada). All three attributes have the same importance according to the need of organization and it would be unethical to isolate any of these attributes from the information security training and awareness program (Ally, 2015). There are some other motivational factors that encourage this research to be carried out and are briefly discussed below:

i. Improper management:

When there is no support from organization management, information security training and awareness program are always unsuccessful. Moreover, there is ignorance by senior managers regarding information security training and awareness program and also there is no specific budget for those programs. In some organizations, the management isolates the need of organization and implements the training and awareness session. This kind of practice of management believed that organization has implemented the best training and awareness program but does not deliver the desired result and blaming either employees or the training and awareness program organiser (Chelsa, 2002).

ii. The inappropriate way of delivering the message:

Delivering the message and information about training and awareness program plays a vital role in success of the program. Many researchers and organizations believed that online videos, presentation and lectures are the effective tools for training and awareness program, but online presentation and videos are often ignored by the users and classroom lectures are boring. Computer-based training method and instructor based training method are two types of training method, but the user should be given freedom to choose the training method otherwise rather than focusing the training session user will be more focusing on completing the training session and getting the completion certificate.

iii. Implementing the incorrect metrics to measure the efficiency:

Prior to implementing the information security training and awareness program, the three factors need to be considered: what needs to be measured; how to measure and when to measure? Which will be the correct methodology for data collection while measuring the user's knowledge, whether the survey will be the best method for data collection or the interview (Schroeter, 2014)? It has been also stated that one survey questions or an interview question do not decide or measure the knowledge, attitude and performance of people (Chen, Li, Hoang, & Lou, 2013). Which statistical theory best suit for analysing the collected data? How frequently or what would be the time interval for collecting the data again for the periodic review of measuring the efficiency of information security training and awareness program (Kahraman)?

3. Motivation

Employees are considered as the biggest threats to an organization and are the main reason for security infringement. Prior to implementing information security training and awareness program, organizations must discuss with organizational users regarding data and information security and what benefits will be to the organization and its users after implementing the information security training and awareness program (Herold, 2010). Highly trained and aware employees are always the basis of secure and reliable that every organization is insisting. So implementing effective and worthy information security training and awareness program helps in improving the employee's performance, which at the end improves the performance of an organization (Kadel, 2004). Generally, in most of the organizations, it is believed that the information technology department is responsible for any information security incidents (Karjalainen, 2011). But this paper clarifies the misunderstanding that the information technology department of an organization is only responsible for information security incidents and other department and employees are isolated from the security incidents, but it is every employee role and responsibility to address those incidents (Allen, 2013).

The initial purpose of the information security education, training and awareness program is to give away a clear message that employee is familiar with their own terms (Hight, 2005). Information security training and awareness program must concentrate on how to protect the organization and its asset, why it is necessary to protect that and what happens if fail to protect the organizational information and asset (Knapp & Ferrante, 2015). Information security training and awareness program should draw user's interest on organization's information security policy and procedure which helps in developing the information security culture in an organization and reaffirming users about information security policy, its results and user's responsibility in security policy (Khan, Alghathbar, Nabi, & Khan, 2011).

The main purpose of this paper is to investigate and measure the efficiency of information security training and awareness program. This paper helps in developing the continuous steps in judging the employees in an organization which is possible only through measuring the efficiency of information security training and awareness program (Lavrakas, 2010). Organizational users who are directly involved in information security system are more probable advantageous from this paper (Elkhannoubi & Belaissaoui, 2016). Thus, investing in information security training and awareness program indicates that there will be a probability of less information security incidents and make certain that the business status is operating as anticipated (Safianu & Twum, 2016).

Based on the above scenario, this paper addresses the following research problems.

How measuring Information Security Education Training and Awareness framework is effective in minimizing the organizational information security incidents?

To achieve the best solution the research question has been divided into following sub research questions are listed below:

1. How information security training and awareness framework are evaluated and its success factor is measured?
2. How employee's behaviour affects the overall information security in an organization?
3. How employee's behaviour will be changed after implementing information security training and awareness program?

4. Related work

The literature review involves in recent and previous literature, investigate and analysis for verification. The current literature review comprises of information security training and awareness factors and explaining the employee's information security related incidents and its outcomes. The literature review concentrates on re-evaluating the confidentiality, integrity and availability of information security attributes. This literature also re-examines the information security training and awareness program and organization's information security policy from an employee's position. This literature review is inherent from the limitation of previous research which instigates the source for this paper. This literature review rationally explores the relationship between organizational information security incidents, employee's behaviour on organizational information security and the necessity of information security education, training and awareness program (Poepjes & Lane, 2012).

According to Nurul (Molok, Chang, & Ahmad), organizations are experiencing more information security incidents from organizations employee's negligence and lack of knowledge, but this lacks the study of information security incidents from external attack and management point of view. Agata (McCormac, Parsons, & Butavicius, 2012) agrees that investing in information security is the investment in technological aspects and training and awareness session as well. Hekkala (Hekkala, Vayrynen, & Wiander, 2015) insists that information security professionals are only responsible for information security incidents in an organization, but by saying so do not signify that other employees are isolated from the information security incidents. Information security incidents in an organization do not only disrupt the organization's information system but also compromises the employee's privacy and security. Straub and Welke (Straub & Welke, 1998) agree that information security training and awareness program are frequently ignored by the organizational management and even by employees as well. Information security training and awareness program should not only focus on educating and changing the behaviour of employees but also should focus on its return on investment in the organizational business. Stephen (Stephanou & Dagada) agrees that most of the information security incidents arise in an organization from employee's lack of knowledge. To address this issue, the organization must implement information security training and awareness program and make it part of organizational information security policy. The outcomes dramatically reduce the information security related incidents by half in the first phase, but this is not the total solution as information security training and awareness is a continual process.

According to Kreicberga (Kreicberga, 2010), organizations are implementing the information security training and awareness program without realizing the need of organization and also without knowing the employee's level of understanding. In addition to this, the organization implements the inappropriate metrics to measure the efficiency which results in zero outcomes. Therefore, this literature addresses all of the above-mentioned issues and attempts to fulfil the gaps that have been addressed.

5. Methodology

5.1 Data collection

The survey was carried out online through survey monkey and the participants were mostly from the United States. The required candidates for this survey were around 100 but only 55 responded. Among them, approximately 65% were above the age of 45 and the remaining were in the range of 18- 45. In comparison to gender, female participants were more than a male. The Quantitative research method was implemented as a research methodology because the survey was considered as a data collection method. The survey questions were designed to measure employee's knowledge and behaviour and how employees respond to the information security incidents. The survey questions were categorised into three different sets: first set of survey questions for Information Technology professionals, second set for all employees and management and third of the control group (here control group refers to the group of employees who are participating in the data collection but they would not be receiving any kind of training and awareness program). Survey questions were prepared and were approved by the principal supervisor and co-supervisor for the verification. In every set of questions employee's knowledge and behaviour in relation to information security is measured. The questions were multiple choices and one more option was added in case if participants would like to express their own opinion. The survey questions that were asked to participants were authorized and certified by qualified and knowledgeable experts in the area of social science research as well as information technology security professionals. All three set of survey questions were handover to participants on the same day and allowed two days to complete the survey.

5.2 Research process

The initial survey questions were designed to know the employee's level of understanding regarding the information security education, training and awareness and information security policy. After the survey questions were designed then the survey questions were uploaded into survey monkey for the survey. The requested amount of participants was 100 but only 55 responded and the survey was closed by survey monkey team after 2 days. Closed-ended questions were asked in this group and one more option was added if participants want to express their own opinion. The participants can choose the options from three multiple choices and if they have any opinion in regards to the survey question then they can select the additional option and write down their opinion. The authenticity as apparent by the participants needs to be discussed in terms of meaningful replies in relation to the component of the area of the research participants were enquired about. The response requires various sources of participant's intelligence which also involves in observing their values, cultures and the method they explain and know this.

The second set of questions was designed after analysing the result of the initial survey. In the initial survey participants level of knowledge and understanding in relation to information security was measured. Based on those result the survey questions were designed. If the data was not collected through online mode then there needs to implement some sort of information security training and awareness program to the participants but it was not possible in the online method of data collection. Therefore, there is a need to implement new techniques for conducting information security training and awareness session, which includes some useful and real-time scenario for the participants in every question and participants, has to respond based on those scenarios. The purpose of introducing the scenario is to make participants aware of the information security policy and information security education training and awareness program. The second survey questions will be implemented after analyzing the result of the initial survey. The result of this survey will be compared with the initial survey and result will be analyzed.

The third set of questions will be same as the second set, but the only difference is that there will be no scenario. The survey will be conducted after a couple of month's time interval after conducting the second survey. The purpose of this survey is to measure whether participants are keeping in mind the information security policy and applying it in the workplace or is participating closing the eyes to information security policy and procedure. This survey also measures the number of information security incidents that had happened in the past and have occurred recently.

This research also surveys another group known as the control group. In the control group, participants will be invited for the initial survey and final survey within the time interval of 3 months, but the participants would not be participating in any information security related training and awareness program. These survey results were then compared to survey results from the other two groups who received the information security training and awareness. The purpose of this comparison is to know whether changes in organizational security culture forces or motivates the employees to change their behaviour or not. If the change in the organizational culture motivates the employees for changing their behaviour by sticking with the organizational information security policy then there will be the minor necessity of information security education training and awareness program.

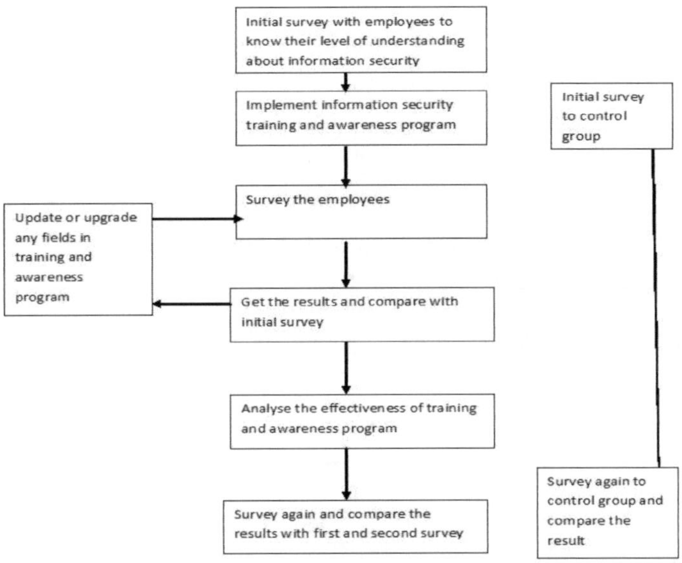

Figure 1: Data collection process

6. Data Analysis

6.1 Initial survey analysis:

ANOVA test and F-ratio test were implemented in this research for comparing the independent means of the survey grades and marks of the participated employees.

Null hypothesis H0: while comparing the results and conclusion for an initial survey of participants prior to participating in information security training and participants from control groups, there will be no major variation statistically.

Alternative hypothesis Ḩ: while comparing the results and conclusion for an initial survey of participants prior to participating in information security training and participants from control groups, there will be major variation statistically.

Table 1: Means of initial survey of both group

	N	Mean	Standard Deviation	Standard Error	95%confidence interval for Mean	95%confidence interval for Mean
					Lower bound	Upper bound
1	55	7.69	1.015	0.155	7.40	8.04
2	55	8.26	1.130	0.164	7.95	8.60
Total	110	8.00	1.112	0.118	7.78	8.24

Here both group means the first group who will receive the information security training and awareness program and a second group who won't receive any sort of information security training and awareness(Kim & Homan, 2012) (Zaiontz, 2017).

Table 2: F-ratio of initial survey of both group

	Sum of squares	Df	Mean Square	F	Significance
In between groups	7.105	1	7.105	6.050	0.015
Within the groups	98.785	84	1.171		
Total	105.89	85			

In the initial survey, there is substantial variation statistically in both groups at 0.015 level of significance. The F ratio value is equal to 6.050 with the significance level of 0.015 which is less than the critical value, accepts the alternative hypothesis (Kim & Homan, 2012) (Zaiontz, 2017).

6.2 Final survey analysis:

Null hypothesis H0: while comparing the results and conclusion of participants who participated in information security training and participants who did not take part in information security training and awareness program, there will be major variation statistically.

Alternative hypothesis H: : while comparing the results and conclusion of participants who participated in information security training and participants who did not take part in information security training and awareness program, there will be no major variation statistically.

Table 3: Means of final survey of both group

	N	Mean	Standard Deviation	Standard Error	95%confidence interval for	95%confidence interval for

					Mean	Mean
					Lower bound	Upper bound
1	55	7.95	1.041	0.180	7.59	8.32
2	55	8.06	0.800	0.108	7.85	8.30
Total	110	8.03	0.896	0.0969	7.84	8.21

Table 4: F-ratio of final survey of both group

	Sum of squares	Df	Mean Square	F	Significance
In between groups	0.305	1	0.308	0.374	0.540
Within the groups	69.500	84	0.816		
Total	69.805	85			

The table illustrates that there is no major variation statistically in between the participants who participated in information security training and participants who did not take part in information security training and awareness program scores at 0.05 level of significance. The significance level of 0.54 was equal to the value of F-ratio 0.374 and the critical value is not more that level of significance the null hypothesis could not be rejected.

6.3 Efficiency in between training methods

In order to discover the efficiency in between the training methods, five training methods have been selected 1: Instructor based training method 2: Computer-based training method 3: Group discussion 4: Online discussion forum and 5: Another method. Participants were asked to choose one of five training methods. Below is the chart that illustrates which the best training method is.

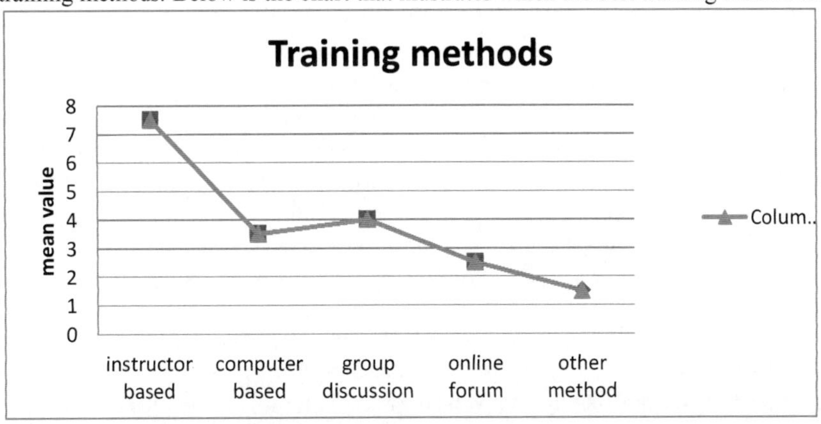

Figure 2: Chart showing the various training methods

7. Finding and discussions

It is clear in this paper that information security training and awareness program is proficient in changing the knowledge, behaviour and attitude of the employees in an optimistic manner. While discussing the training mode, there is not a major difference in training mode according to participant's responses. Despite the benefits and limitations of training methods, organization management should consult with participants, which training method they want rather than forcing them for one method. The organization should look after participant's convenience in delivering training method rather than management because, at the end, participants need to be beneficial for the training. This paper also reveals that participants are now more concerned about information security and organizational information security policy. Participants now believe that it is also their responsibility to ensure the safety and security of organizational information system and also participants are motivated to report any information security incidents. The outcomes further explain that organizational management should play a significant role in either continuing the training and awareness program or implement one-time training and awareness program and whether outsourcing the training and awareness program or create one by consulting with information security professionals. The outcomes also make clear that investment in information security training and awareness program assists in gaining valuable revenue in the long term by increasing the return on investment.

This paper contradicts that one question in the survey does not necessarily measure the performance of the employees. For the success of information security training and awareness program, organization management should ensure that all participants understands the organization information security policy and procedure and apply them in the workplace. But still, how is management measure that participant understands the training and awareness program? One of the effective measures is that let the organizational management be the listener and current participant be the trainer for the new participant. This method assists management in measuring how current participants present the awareness session to the new participants? Has the current participant fully known about the organization's information security policy and procedure and is he/she is being able to make it clear to the new participants about the importance of the organization's information security policy and procedure?

This paper also explores that information security training and awareness program should be reviewed in order to maintain its effectiveness. Prior to implementing the information security training and awareness program did the organization cross investigate the program with other organizations? Was the program successful in that organization? Did the program increase the ROI (return on investment) and changed the performance of organization's security culture? The purpose of reviewing the information security training and awareness program is to analyse whether the training and awareness program that was implemented achieved the goal. Furthermore, the reviewing process involves in critical thinking of what has been succeeded and how to make progress in the future.

It has been stated that successful information security training and awareness program is proficient in changing employee's knowledge, performance and attitude which makes an optimistic transformation in organization security culture, but this is not the only reason for successful information security training and awareness program. There are other factors that need to be investigated and analysed further. The collected data should be analysed through other statistical methods for reliability and validity of the study.

Further study should be carried out on how age factor affects in the training methods. Separate analysis should focus on why the younger generation is much interested in computer or web based training rather than instructor based training and group discussion and why middle age group participants are more focusing on the instructor based training method and group discussion. While analysing the training methods, age factor plays a significant role. Young generations prefer computer based training method while middle age group prefers the instruction based and group discussion based training method. The capable of critical thinking, understanding and quick learning is high in the young generation and this feature makes young employees to adjust in computer based training method rather than sitting in lecture for hours. Moreover, young generation is quick to adapt to change and technology is changing. But some organizations could say that it is not necessarily important that age factor affects on the training methods. It also depends on the organizational culture and environment and the employee background as well. For example, it would be beneficial for information security professionals to implement the computer based training methods as they are familiar with the information security terms and acronym. For the normal employee, instructor based training methods would be advantageous because there needs to be someone who can explain any technical words and terms and group discussion would be beneficial for the management people as they can discuss about the problem and find out the right solution. This also assists them in decision making in the future by sharing the experiences and ideas during the group discussion.

As the technology is growing the new threats and challenges to information security is also emerging. Therefore, further study is needed to investigate on new threats and challenges. Rather than blaming the employees, organizations should find out the weakest point of the organization through penetrating testing and ethical hacking. Also,organization should perform phishing attacks, and social engineering to the employees to know how employees react to those incidents. Further investigating needs to be conducted on phishing attacks and penetrating testing.

8. Conclusion

In conclusion, this research paper had examined the research questions of information security awareness and training plan with the purpose of adding up the theory and procedures. This research paper also explores the various types of outlook that are implemented by information security researchers and professionals concerning these essential features of security awareness. This research paper also explores that implementation of security policy and measuring the awareness level in employees. This research paper addresses the gaps that have been identified in the literature review. This paper also addresses the problem which encourages for this research. Therefore, this paper verifies that it would beneficial for the organizational users who are beginning and new to organizational information security system.

References

Allen, J. H. (2013, 5 13). Security Is Not Just a Technical Issue. Accessed via https://www.us-cert.gov/bsi/articles/best-practices/governance-and-management/security-is-not-just-a-technical-issue on 11 October 2017.

Ally, S. (2015, 03 25). Three dimensions of information systems. Accessed via www.slideshare.net/suleymans19/three-dimensions-of-information-systems on 11 October 2017

Braue, D. (2014, 5 7). Cost of Australian IT security incidents rises despite improving information sharing. Accessed via www.cso.com.au/article/544481/_cost_australian_it_security_incidents_rises_despite_improving_information_sharing/ on 6 October 2017

Brodie, C. (2008, 6 30). The Importance of Security Awareness Training. Accessed via www.sans.org/reading-room/whitepapers/awareness/importance-security-awareness-training-33013 on 9 Ocotber 2017

Chelsa, R. (2002, 10 25). Security Awareness Implementing an Effective Strategy. Accessed viawww.sans.org/reading-room/whitepapers/awareness/security-awareness-implementing-effective-strategy-418 on 13 October 2017

Chen, H., Li, J., Hoang, T., & Lou, X. (2013). Security Challenges of BYOD: a Security Education, Training and Awareness perspective. Accessed via minervaccess.unimelb.edu.au/bitstream/handle/11343/33347/300316_2013_Li_SETA.pdf?sequence=1 on 10 October 2017

Edwards, M. (2015, 4 15). Identity theft: More than 770,000 Australians victims in past year. Accessed via www.abc.net.au/news/2015-04-14/identity-theft-hits-australians-veda/6390570 on 5 October 2017

Elkhannoubi, H., & Belaissaoui, M. (2016, 10). User's Behaviors Influence on Cybersecurity Strategy Effectiveness. Accessed via www.researchgate.net/publication/309560700_User's_Behaviors_Influence_on_Cybersecurity_Strategy_Effectiveness on 10 October 2017

Giovanni. (2015). 10 Reasons Why Employees Need Security Awareness Training. Accessed via www.expertsecuritytips.com/employees-security-education-awareness-training on 13 October 2017

Hekkala, R., Vayrynen, K., & Wiander, T. (2015). Nordic Contributions in IS Research. 6th Scandinavian Conference on Information System,SCIS 2015 , 5-50.

Herold, R. (2010). Why Information Security Training and Awareness Are Important. Accessed via www.infosectoday.com/Articles/Security_Awareness_Training.htm 11 October 2017

Hight, S. D. (2005). The importance of a security, education, training and awareness program (November 2005). Accessed via www.infosecwriters.com/Papers/SHight_SETA.pdf on 9 October 2017

Identity Theft. (2016, 4 19). Australian Bureau of Statistics: Accessed via www.abs.gov.au/ausstats/abs@.nsf/0/90A4971FB1FC0992CA2579E40012060A?opendocument on 5 October 2017

Kadel, L. A. (2004, 3 24). Designing And Implementing An Effective Information Security Program: Protecting The Data Assets Of Individuals, Small And Large Businesses. Accessed via www.sans.org/reading-room/whitepapers/hsoffice/designing-implementing-effective-information-security-program-protecting-data-assets-of-1398 on 8 October 2017

Kahraman, E. (n.d.). Evaluating IT security performance with quantifiable metrics. Accessed via pdfs.semanticscholar.org/1a28/bab7edab76c234984dd9a3c96eb76d661d3b.pdf on 10 October 2017

Karjalainen, M. (2011, 10 28). Improving employee's information systems (IS) security behaviour: Toward a meta-theory of IS security training and a new. Accessed via jultika.oulu.fi/files/isbn9789514295676.pdf on 9 October 2017

Khan, B., Alghathbar, K. S., Nabi, S. I., & Khan, M. K. (2011). Effectiveness of information security awareness methods based on psychological theories. African Journal of Business Management Vol. 5(26) , 10862-10868.

Kim, P., & Homan, J. V. (2012). Measuring the effectiveness of information security training: A. Issues in Information Systems, 13 (1), 215-224.

Knapp, K. J., & Ferrante, C. J. (2015). Policy Awareness, Enforcement and Maintenance: Critical to Information Security Effectiveness in Organizations . Journal of Management Policy and Practice vol. 13(5) , 66-80.

Kreicberga, L. (2010). Internal threat to Information security. Accessed via epubl.ltu.se/1653-0187/2010/050/LTU-PB-EX-10050-SE.pdf on 11 October 2017

Lavrakas, P. J. (2010, 5). An Evaluation of Methods Used to Assess the Effectiveness of Advertising on the Internet. Accessed via www.iab.net/media/file/Evaluation_of_Internet_Ad_Effectiveness_Research_Methods.pdf on 9 October 2017

Maqousi, A., Balikhina, T., & Mackay, M. (2013). An effective method for information security awareness raising initiatives. International Journal of Computer Science & Information Technology (IJCSIT) Vol 5, No 2 , 63-72.

McCormac, A., Parsons, K., & Butavicius, M. (2012, 4). Preventing and Profiling Malicious Insider Attacks. Accessed via www.dtic.mil/dtic/tr/fulltext/u2/a563808.pdf on 8 October 2017

MeasuringOutcomes. (n.d.). Strengthening non profits: Accessed via http://www.strengtheningnonprofits.org/resources/guidebooks/MeasuringOutcomes.pdf on 12 October 2017

Molok, N. N., Chang, S., & Ahmad, A. (n.d.). Disclosure of organizational information on social media: perspectives from security managers. Accessed via www.pacis-net.org/file/2013/PACIS2013-108.pdf on 8 October 2017

Poepjes, R., & Lane, M. (2012). An Information Security Awareness Capability Model (ISACM). Accessed via http://www.ro.ecu.edu.au/cgi/viewcontent.cgi?article=1136&context=ism on 13 October 2017

Russell, C. (2002, 10 25). Security Awareness - Implementing an Effective Strategy. Accessed viawww.sans.org/reading-room/whitepapers/awareness/security-awareness-implementing-effective-strategy-418 on 5 October 2017

Safianu, O., & Twum, F. (2016, 6). Information System Security Threats and Vulnerabilities: Evaluating the Human Factor in Data Protection. Accessed via www.researchgate.net/publication/304066003_Information_System_Security_Threats_and_Vulnerabilities_Evaluating_the_Human_Factor_in_Data_Protection on 12 october 2017

Schroeter, J. (2014, 2 12). Measuring the effectiveness of your security awareness program. Accessed via www.csoonline.com/article/2134334/metrics-budgets/measuring-the-effectiveness-of-your-security-awareness-program.html on 9 October 2017

Stephanou, A., & Dagada, R. (n.d.). The impact of information security awareness training on information security behaviour: The case for further research. Accessed via www.citeseerx.ist.psu.edu/viewdoc/download?doi=10.1.1.145.1461&rep=rep1&type=pdf on 5 October 2017

Straub, D. W., & Welke, R. J. (1998). Coping with System Risk. Security Planning Models for Management decision MIS Quarterly, Vol. 22, No. 4 , 441-470.

Zaiontz, C. (2017). Two within subjects factors. Accessed via www.real-statistics.com/anova-repeated-measures/two-within-subjects-factors/ on 13 October 2017

Agricultural Environment Monitoring Using Internet of Things and Wireless Sensor Networks

Mohammad Sanaullah Chowdhury, Shreya Khisa and Sourav Kumar Das
University of Chittagong, Chittagong, Bangladesh

E-mail: sana1691@gmail.com, shreyakhisa21@gmail.com, souravcsecu@gmail.com

Abstract

Agriculture is the main food production system of a country. Within the decades there are many research work has been demonstrated to improve the cultivation of agriculture using new technology. In this paper, an agricultural environment monitoring system is described using Internet of Things, open source platform Raspberry Pi and Wireless Sensor Network. In this system, the Wireless Sensors collect the data of various environmental parameters such as temperature, humidity, soil moisture etc. and send the collected data to Raspberry Pi. The Raspberry Pi stores the data in a database of a web server. The data can be monitored from anywhere of the world. A client side website is developed to show the data. If the collected environmental data reaches in a threshold condition, a gmail notification is sent to the user as a warning. In this paper, overall description of the system architecture, used hardware and software components are described in details.

Keywords: Raspberry Pi, Wireless Sensor Network, Environmental monitoring, Internet of Things, Agriculture.

1. INTRODUCTION

Within the decades the use of internet has drastically changed from mainframe computers to smart phones. At the turn of ages, a new concept came across which is called Internet of Things (IoT).Internet of Things(IoT) means interconnected devices of a physical network which can be used to control or sense any device in the network that is used in indoor or outdoor for sharing data and resources. Sensors are used to convert physical quantity into numerical data. The network which is constructed with various wireless sensors is called Wireless Sensor Network (WSN). In recent days, the use of IoT and WSN is emerging widely to solve different problems to ease the day to day life of human. Furthermore, the production of food is a very important factor for survival of human.

Much research has been taken place for improvement of the food production system. Hwang, Jeonghwan (2010) presented a system for environment monitoring for agriculture that collects environmental and soil information of the outdoorthrough WSN-based sensors, image information through CCTVs, and location information using GPS modules. It processes the data using a MSP430 MCU and transmits them to relay nodes and gateways, using a CC2420 RF chip.

Some have presented system to do future prediction of environment. Gaddam,Anuroop (2014) addressed a system of drought monitoring using Wireless Sensor Networks (WSN). It can predict drought situation using collected soil moisture data.

Radio-frequency identification (**RFID**) uses electromagnetic fields to automatically identify and track tags attached to objects. The tags contain electronically stored information. Passive tags collect energy from a nearby **RFID** reader's interrogating radio waves.

RFID technology is also used to in some researches to monitor the environment. As like, Zhao, Ji-Chun (2010), they proposed a structure of agriculture greenhouse production environment measurement and control system based on wired and wireless sensors using RFID technology.
Vasilenko, M. Ulman (2015) presented a concept of horticulture ambient intelligence system that determines watering efficiency in agricultural and horticultural activities.
Sahu (2015),Tarange (2015), Agrawal (2015) proposed an automatic irrigation and controlling system using Raspberry Pi and sensors to optimize the use of water.
Sheikh Ferdoush and Xinrong Li (2014) presented a generalized design of environment monitoring system using WSN technology, Raspberry pi and Arduino.
Lazarescu, Mihai T (2013) emphasized on designing WSN platforms and sensor node design for long term environment monitoring. They gave importance to design low cost system to monitor environment. TongKe, Fan (2013) found integration of cloud computing, RFID technology and IoT useful to create a smart agriculture environment.

2. OVERALL SYSTEM ARCHITECTURE

In Figure 1.the block diagram of the proposed system architecture is shown. The sensor collects data from the environment. The Raspberry Pi processes the collected data and stores it in a database of a web server. The users can access to the data through a web application through browser which has an internet connection in it. The digital sensor can directly communicate with the Raspberry Pi 3 through its GPI/O pins.But Raspberry Pi does not have analogue pins. So, it cannot communicate with the analogue sensors directly. In that case, an analogue to digital converter is used to get digital data from analogue sensors.

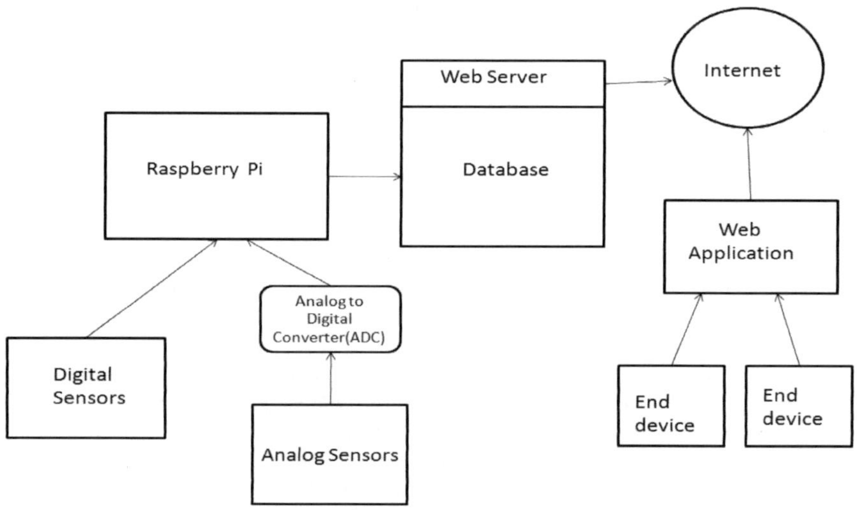

Figure 1. Block diagram of the proposed system architecture

3. HARDWARE PROTOTYPE

3.1. Raspberry Pi 3 Model B

The main system is embedded in Raspberry Pi 3 Model B.Raspberry Pi is a credit card size microcomputer. The Raspberry Pi 3 is the third generation Raspberry Pi.It has 1.2GHz 64-bit quad-core ARMv8 CPU, 802.11n Wireless LAN, and Bluetooth 4.1Bluetooth Low Energy (BLE). It also has 1GB RAM,4 USB ports,40 GPIO pins, Full HDMI port, Ethernet port, combined 3.5mm audio jack and composite video, Camera interface (CSI), Display interface (DSI),Micro SD card slot (now push-pull rather than push-push)Video Core IV 3D graphics core. It requires 2 Amperes of current to power up and can tolerate up to 5Volts.

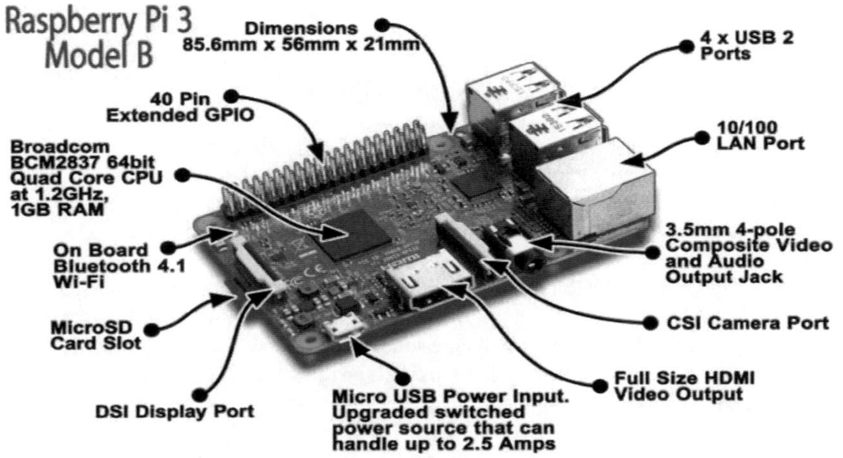

Figure 2. Raspberry Pi 3 Model B

3.2. Waterproof DS18B20 Digital Temperature Sensor

Waterproof DS18B20 Digital Temperature Sensor is used to measure temperature. It is used as a backup for readingtemperature in a wet condition for its waterproof facilities. It can tolerate maximum temperature 125°C but recommended temperature is 100°C.It requires max voltage 5V and min voltage 3.3V.

3.3. Grove Moisture Sensor

To measure soil moisture level, Grove moisture sensor is used.Soil moisture sensor based on soil resistivity measurement. It is very easy use. It has2.0cm width and 6.00cm long two probes which are inserted into the soil to measure the moisture level. The sensor works as with the specifications given in Table 1.

Table 1. Specifications of Grove moisture sensor

Item	Condition	Min	Typical	Max	Unit
Voltage	-	3.3	/	5.00	V
Current	-	0	/	35	mA
Output Value	Sensor in dry soil	0	-	300	/
	Sensor in humid soil	300	-	700	/
	Sensor in water	700	-	950	/

3.4. Analogue to Digital Converter

The ADS1115 is a great analogue to digital converter that iseasy to use with the Raspberry Pi using its I2C communication bus. The ADS1115 is a higher precision 16-bit ADC with 4channels. It has a programmable gain from 2/3x to 16x so it can amplify small signals and read them with higher precision. It is used in the system to convert the sensor's analogue data to digital data.

4. PROPOSED SYSTEM AND IMPLEMENTATION

The main purpose of the proposed system is to monitor the environment of the agricultural field remotely. This is an application of Internet of Things. At first, Raspbian Jessie the latest operating system of Raspberry Pi, which is based on Debian Linux, is installed in the 8GB SD card of the Pi.As Raspberry Pi 3 supports Wireless LAN, it is connected wirelessly with the internet.

Raspbian Jessie directly boot with GUI interface which is an advantage of using it than earlier versions of Raspbian. The Raspbian is updated and upgraded to the newest features. The two sensors used for the system works with two different types of interfaces. The temperature sensor works with 1-wire interface and ADS1115 4-channel ADC which is needed for moisture sensor works with I2C interface. Both interfaces are enabled using the command in the terminal "sudoraspi-config".

Grove Moisture sensor is an analog sensor it cannot give direct output to Raspberry Pi as Pi does not have any analog GPIO pins. In that case, it needs an Analog to Digital converterto give digital output. ADS1115 4-channel ADC works here. At first, the ADC is connected with the Raspberry Pi. In the secondstage, moisture sensor is connected with Pi and the SIG pin of the sensor is connected with one of the channel of the ADC.Wiring of the Raspberry Pi GPIO pins with the sensors and the ADC is shown in Figure 3.

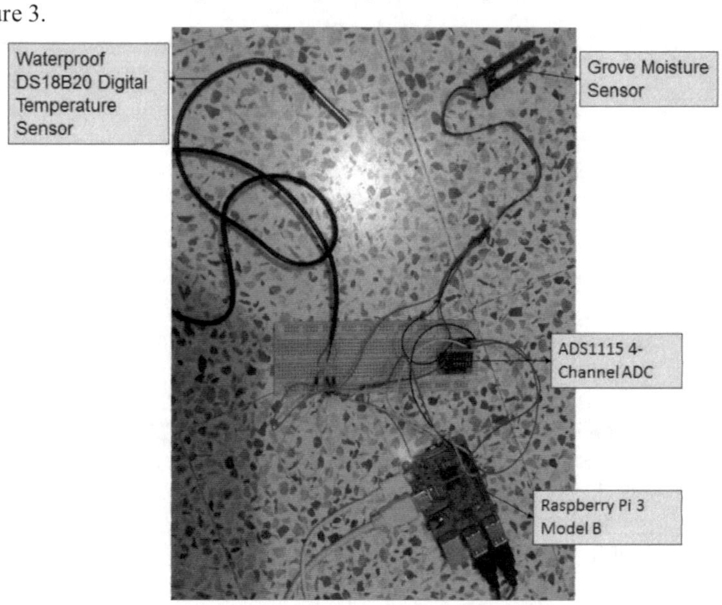

Figure 3. Final circuit design of the system

The advantage of using Raspberry Pi is that it can be configured with static IP address. In our system, Raspberry Pi 3 is configured with a static IP address by editing the file with the command in the terminal "sudonano /etc/dhcpcd.conf ". With a static IP address, remote access of the Pi is much easier than having a dynamic IP address. It is configured with the IP address 192.168.1.21.

MySQL database is used to store the collected data by the sensors. It is a relational database management system. To make the development cost low, Apache2 Web Server is used as a server of the database. It is a local host web server. But the system can be easily transferred to a global hosting web server when necessary.Two protocols are used in the development of the system.SSH (Secure Shell) Protocol is used to access the Raspberry Pi remotely.This is also enabled using the "sudoraspi-config" command in the terminal.SMTP (Simple Mail Transfer Protocol) is used to send gmail from Pi to user when a threshold level is reached. The SMTP protocol is enabled in the program explicitly using the SMTP provider with port 587.

The client-side interface is implemented with HTML, CSS, PHP, crontab. HTML and CSS is used in combination to mark up and style the web page. PHP is used to create the back-end, which is server side. It established connection between the database and the client side website.**cron** is the system processwhich will automatically perform tasks according to a schedule. The schedule is called the **crontab**. It is a service of Linux operating system. In our system, this service is used to execute our programs after 5-minute of intervalto collect temperature and soil moisture data.

5. EXPERIMENTED RESULTS

The data collected by the temperature sensor is taken by an experiment at home in a day. The results are found by sensing the temperature with thetemperature sensor is shown in Figure 4.

ⓘ 192.168.1.21/temperature_n.php

Date	Temperature
2017-06-02 18:35:03	29.40
2017-06-02 18:40:02	29.40
2017-06-02 18:45:02	29.40
2017-06-02 18:50:02	29.40
2017-06-02 18:55:02	29.40
2017-06-02 19:00:02	29.40
2017-06-02 19:05:02	29.40

Figure 4. Experimented Temperature data fetched by temperature sensor in the same day with 5-minute interval

Moisture sensor data is taken by an experiment using a flower pot with dry,humid andwater in soil. First the soil was dry, after giving a little water, it became humid, and after giving a lot of water, it became watery. The experimented data collected from the sensor is shown in Figure 5.

192.168.1.21/moisture.php

Date	Moisture Level	Moisture Condition
2017-06-02 10:25:41	0.00	Dry soil
2017-06-02 10:33:29	722.00	Water in soil
2017-06-02 10:33:45	0.00	Dry soil
2017-06-02 10:44:11	0.00	Dry soil
2017-06-02 10:45:11	72.00	Dry soil
2017-06-02 10:46:02	823.00	Water in soil
2017-06-02 10:46:22	0.00	Dry soil
2017-06-02 10:48:34	719.00	Water in soil
2017-06-02 10:50:01	726.00	Water in soil
2017-06-02 10:50:10	651.00	Humid soil

Figure 5. Experimented Moisture data fetched by moisture sensor

6. SUMMARY AND CONCLUSION

A web interface is successfully developed to collect data from a system based on Internetof Things(IoT) and Wireless Sensor Network (WSN). This system can be used in a remote area and it can serve as a benefit to the farmers.They do not need to go to their field regularly to check the weather. They can easily monitor the weather if they live far away from the fields. It can help them to concentrate on other works outside of their farming. The hardware and softwareused in this system is cheaper thanother environment monitoring system. The price of Raspberry pi 3 is very low and is about only $35. It is also environmentfriendly as the power consumption of Raspberry Pi is very low. It can be deployed at an outdoor location for a long term basis. As all the software and programming tools which are used also open source, it keeps the cost of the development very low. As a future work, system proposed here can be expanded in different ways.Security in IoT is a great concern today as the IoT devices are increasing rapidly.Alsoenormous data problem will also arise when the system will be deployed in a large area in real life.Taking these accounts, in near future,we would like to extend this system with security so that no authorized access can be made and cloud storage for the system to make the systemmore scalable in future deployment. Taking data from the sensors, some statistical analysis can be madeto forecast the weather for particular area or proper time of cultivation of crops like paddy, corn etc.

REFERENCES

Hwang, Jeonghwan, Changsun Shin, and Hyun Yoe (2010). Study on an agricultural environment monitoring server system using wireless sensor networks, *Sensors* 10.12 (2010): 11189-11211.

Gaddam, Anuroop (2014). Designing a wireless sensors network for monitoring and predicting droughts (2014).

Zhao, Ji-chun (2010). The study and application of the IOT technology in agriculture, *Computer Science and Information Technology (ICCSIT), 2010 3rd IEEE International Conference on*. Vol. 2. IEEE, 2010.

Vasilenko, Alexandr, and MilošUlman (2015).Concept of Horticulture Ambient Intelligence System, *AGRIS on-line Papers in Economics and Informatics* 7.4 (2015): 209.

Sahu, Chandan Kumar, and PramiteeBehera (2015). A low cost smart irrigation control system, Electronics and Communication Systems (ICECS), 2015 2nd International Conference on. IEEE, 2015.

Tarange, Pandurang H., Rajan G. Mevekari, and Prashant A. Shinde (2015).Web based automatic irrigation system using wireless sensor network and embedded Linux board,*Circuit, Power and Computing Technologies (ICCPCT), 2015 International Conference on*. IEEE, 2015.

Agrawal, Nikhil, and SmitaSinghal(2015).Smart drip irrigation system using raspberry pi and arduino, *Computing, Communication & Automation (ICCCA), 2015 International Conference on*. IEEE, 2015.

Ferdoush, Sheikh, and Xinrong Li (2014).Wireless sensor network system design using Raspberry Pi and Arduino for environmental monitoring applications, *Procedia Computer Science* 34 (2014): 103-110.

Lazarescu, Mihai T(2013). Design of a WSN platform for long-term environmental monitoring for IoT applications,*IEEE Journal on Emerging and Selected Topics in Circuits and Systems* 3.1 (2013): 45-54.

TongKe, Fan (2013). Smart agriculture based on cloud computing and IOT,*Journal of Convergence Information Technology* 8.2 (2013).

Proc. 1st International Conference on Machine Learning and Data Engineering (iCMLDE2017) 20-22 Nov 2017, Sydney, Australia

DSTUS: A Deep Learning approach for Smart Traffic Updating System in GIS

Minhaz Uddin Ahmed[1], Abolghasem Sadeghi-Niaraki [2, 3*], Md. Rezaul Bashar[4], Soo-Mi Choi[3]

[1]Department of Computer Engineering, Inha University, Incheon, South Korea
[2]Geoinformation Tech. Center of Excellence, Faculty of Geomatics, K.N.Toosi Univ. of Tech., Tehran, Iran
[3]Department of Computer Science and Engineering, Sejong University, Seoul, Republic of Korea
[4]Science, Technology and Management Crest, Sydney, Australia
*Corresponding author's E-mail: a.sadeghi.ni@gmail.com

Abstract

Real-time traffic information plays an important role in our daily life. Traffic information helps us to understand the traffic congestion situation of different part of the city for various reasons such as traffic flow of a particular area, exact travel time to a destination, road condition, accident information etc. These days many vehicles carry an electronic device called black box to facilitate the personnel who operate the vehicle. Specifically, to find out the reason behind incident for an insurance claim and correct decision making. In this paper, we proposed a deep learning based smart traffic updating system which exploits black box video data. We collect the black box video data, process them and train them using deep learning model for accurate vehicle detection. After vehicle detection and counting the number, we send traffic information to the server for better decision making such as avoid gridlock situation or incident. Experiment result shows that our proposed method is a very cost-effective way to manage and update GIS information with cutting age technology. A Deep learning based GIS database will ease the needs for traffic information related to the perception and decision-making of transportation. DSTUS system will assist to predict model of the travel time of each road section in a specific city.

Keywords: GIS, CNN, Deep Learning, Big Data, Vehicle Black Box.

1. INTRODUCTION

Geographic information systems for transportation (GIS-T) references the application of geographic information technologies to control transportation systems as well as resolve shipment related challenges. GIS-T plays a pivotal role in our lives in diverse areas such as social, economic, geographic by Michael f. Goodchild (2000). It is a highly correlated with GIS applications where much costs of producing data and making decisions. Transportation is greatly related to the geographical information system to manage, visualize and analyze geographic data in order to improve correctness. In recent years, GIS data become part of GIS-T and helping the traffic management system due to the increasing demand, diverse transportation, and accurate information.

Traditional vehicles are converted to smart cars with carrying numerous devices which collect different types of data such as road information, maps, traffic flow, traffic congestion, traffic pattern, and car black box data. All these data are high in demand to provide state of the art quick solution to resolve the ever-increasing challenges. Gridlock and other security issue become more and more significant every day. Old-fashioned traffic management system is unable to solve this growing requirement by Jean-ClaudeThill (2000). Because traditional traffic application data cannot share data with other domains. Simultaneously, rapid urbanization also playing as a catalyst for environmental pollution as well as danger for the sustainable traffic management system. The regular traffic management system failed to provide smart decision making based on the gathered data. Therefore, GIS database for transportation (GIS-T) is important Liu R., Servières M. and Moreau g (2012). Traffic information among other data in GIS-T database is critical since it leads to very accurate result for any spatial analysis such as route finding in a road network. Therefore, these days accessing to update traffic information is so valuable by G. Salvi (2014).

Most techniques for producing updates traffic information have some limitations. Some techniques are so expensive such as installing traffic capture camera in each segment of the road, especially for developing countries. Some methods are so complex for calculating of various traffic big data. So we need to look forward to finding easy and cheapest methods in Yushi Chen, Zhouhan Lin, Xing Zhao, Gang Wang, and Yanfeng Gu, Member (2014).

These days almost vehicles contain a device called black box which captures video data utilize for many purposes, such as tracking the particular event or measure traffic situation in an area. To receive up to date GIS data, we can apply deep learning technique to improve the overall performance of large volume of black box data.

The objective of this paper is to propose a deep learning based method for updating traffic information for GIS database for a GIS-T application. In the next section, we analyze the literature of GIS-based traffic management system in videos data. We describe our method Deep Learning approach for Smart Traffic Updating System (DSTUS) in section 3. In section 4 we describe experiments carried out based on the black box dataset. In section 5 we explained our experiment result and contribution.

2. RELATED WORK

There have been many types of research done by professionals related to various techniques for updating GIS databases especially traffic data. For example, in general usage by Liu, R., Servières, M., & Moreau, G. (2012) and Kadri-Dahmani, H., & Osmani, A. (2003) presents a method for updating Geographic Information System (GIS) data from video. Alkan, M., & Jacobsen, K. (2015), Champion, N., Stamon, G., & Deseilligny, M. P. (2009) and Weis, M., Müller, S., Liedtke, C. E., & Pahl, M. (2005) explained using satellite Imagery for updating GIS database. Also for specific utilizing base on traffic data of GIS databases by Choosumrong, S., & Raghavan, V. (2011) states modified algorithm to searches of shortest path, time and safety route that supports dynamic changes information to develop a routing algorithm used a real-time cost updates based on current traffic condition for solving of the real road network condition. Road Traffic Estimation from GPS Data Using Incremental Weighted Update was introduced in Sananmongkhonchai, S., Tangamchit, P., & Pongpaibool, P. (2008) and Li, C., Liu, Y., Wang, Q., & Tai, Y. (2008) discussed a model for updating the trajectories of moving objects when unexpected traffic conditions occur.

For this, there are now a growing number of studies and scholarly papers on using Vehicle Black Box (VBB) for various topics. These days more attention is paid to preventing automobile-related violations. That has led to a growing number of VBB by Baker, W. E., & Sinkula, J. M. (2002). Along with VBB simple recording, now evolving into an intelligent one that can act a useful tool to provide convenience for drivers. A VBB system acts as the flight recorder of a vehicle which utilized to record the behavior of a running vehicle as a vehicular digital video recorder system as explained in Lin, C. C., & Wang, M. S. (2010). Most studies applied VBB for the field of accident management and crash data gathering intend to assist drivers to drive safe, and to identify the cause of an accident once it occurs such as Prasad, M. J., Arundathi, S., Anil, N., & Kariyappa, B. S. (2014), Kang, C., & Heo, S. W. (2017), Moon, H. M., Kim, K. H., Lee, M., Kim, P., & Pan, S. B. (2015)

Chung, Y., Song, T. J., & Kim, J. (2017) analyses injury severity in taxi-pedestrian crashes using more accurate crash data by VBBs such as time-to-collision, crash speed, crash angle, and crash region of vehicle and pedestrian. Black box used for proposing a smart vehicle for accident prevention in Mohamedaslam, C., Najeeb, N. A., & Nisi, K. (2016), Chung, Y., & Chang, I. (2015) used to collect crash data obtained from VBB tried to compare and evaluates these type of data by existing road crash data recording method, which has been recorded by police officers based on accident parties' statements or eyewitness's account. Also S. Y. Yi, J. H. Ryu, and C. G. Lee (2010) tried to recognize if a driver is off the lane on a real-time basis using VBB or some time to recognize a dangerous driving style Han, I., & Yang, K. S. (2009).

3. BLACK BOX DATA ANALYSIS BY CNN

A black box is usually attached to a mechanical device to monitor the performance of the vehicle. It is an event data recorder equipment that record and store data that is gathered by different sensor and indicators. This device can generate some alarm based on the video data it records and assists the driver. It also ensures the reason of accident such as driver's awareness, failure of an equipment in a vehicle, natural disaster, and safety issues by Lin, C. C., & Wang, M. S. (2010). Efficient use of this device can help proper maintenance due to its communication ability with the network and alarm system. Black box collects data from external environment using different sensors and camera. If an accident occurs black box can transmit emergency signal for quick emergency rescue by Chung, Y., & Chang, I. (2015). Moreover, the black box provides crucial information for the vehicle lifecycle management. The region-based convolutional Neural Network method R-CNN by Ross Girshick (2015) based object detector applied which achieved excellent vehicle detection accuracy. The R-CNN detector first fine-tunes Convent features then replace the softmax classifier learn by fine-tuning. In each bounding box, regressor training and features are extracted from each object proposals. Black box video data's are ConvNet forward pass for each vehicle proposal. Features are extracted for a proposal by max-pooling the portion of the feature map but performance is sluggish.

Figure 1. Car black box video data

3.1 DEEP LEARNING

Deep learning is a kind of machine learning and more powerful than shallow networks. DL dramatically improve the performance in many different research areas such as visual object recognition, video, audio, drug discovery, genomics [4] etc. There are different types of Deep learning architectures such as deep neural network, deep belief network, recurrent neural network, a convolutional network which gained outstanding results in computer vision, NLP, machine learning, bioinformatics areas.

These days deep learning applied in many GIS research to find out the patterns from a large volume of data that are useful for accurate traffic updating system. Conventional machine learning algorithm cannot process data in raw format however, DL can tackle this issue. In our experiment, we used VGGNet by Karen Simonyan, Andrea Vedaldi and Andrew Zisserman (2013) which is composed of convolutional layers that perform very small (3x3) filters.

3.2 DSTUS

A vehicle usually includes different type of sensors and indicators which acquires different parameters ensures the condition of that vehicle. These sensor and video data are preserved in memory storage such as a flash card. In our proposed method, we take the input video data then pre-process it. We apply our DSTUS algorithm to detect and count the vehicle mentioned in details in figure 2. After that, we send the data such as warning, vehicle condition, and traffic value to a server for decision making. The server contains the database of previously learned GIS information for particular warning and signal. Based on the traffic information such as a number of vehicles and the path information, analyzing the data and computation, server recommends the correct route for the system. Our proposed system also can indicate whether system faced an accident or not based on some parameter values.

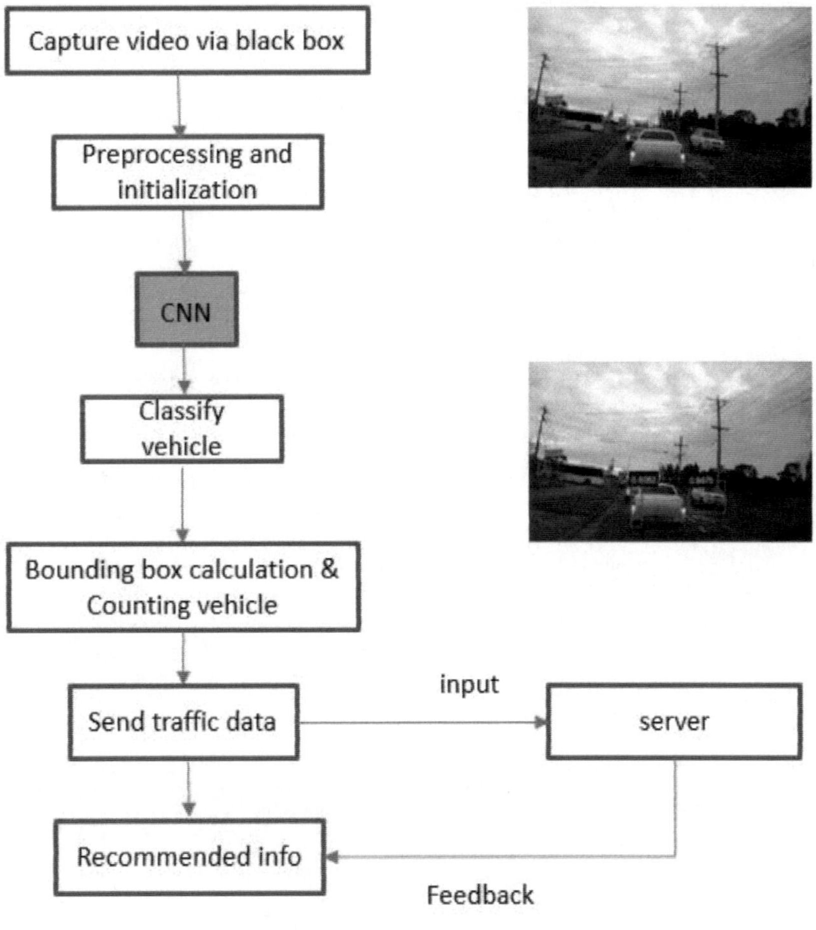

Figure 2. System architecture and design

In a nutshell, the whole flowchart of the proposed DSTUS algorithm as follows.

Algorithm 1. Vehicle Detection and Traffic Updating with DL feature

Input: Captured video clip

Output: Number of vehicle, Update traffic data

Initialize the video reader

For each video frame

Identify regions with intensity values above threshold

Compute CNN features

Classify vehicle

Count total number of vehicle

If number of vehicle > Max

Send traffic data and update

Else If

Continue

End If

End For

4 EXPERIMENTS

In this section, we present our way of experiment with algorithm design. Our proposed algorithm work in several steps. The video reader function reads video file in the first step which provides video properties such as height, weight, frame rate and video format. In the second step consider each frame and convert the color image to grayscale image. We suppress all the objects in the image that are not light-colored cars. In the third step, we apply Fast RCNN by Ross Girshick (2015) to detect the vehicle and remove noise such as small lane marking or other structures. In the fourth step, calculate area and centroid of each object are calculated. Finally, the entire procedure is continued until the N number of frame completed.

Figure 3. Number of car detected and counted based on foreground detector

Figure 3. Shows the number of cars detected from the video frame where the number of the vehicle is two but the foreground detector the only lower portion of the vehicle and counts one.

Figure 4. Number of car detected and total number of vehicle is two.

In figure 4. Shows the total number of the car in a video frame and detects the cars where the detector is based on fast RCNN algorithm. Our experiment uses publicly available VGG net model by Karen Simonyan, Andrea Vedaldi and Andrew Zisserman (2013) that has 5 convolutional layers and 3 fully connected layers. For the vehicle detection, we use Fast RCNN by Ross Girshick (2015) that is convolutional neural network based object detector with deep learning library Caffe. All implementations are done on NVIDIA GeForce GTX 970 graphics card. We use Matlab 2017b as a programming language with matconvnet by Andrea Vedaldi and Karel Lenc (2015) library using Windows platform in our experiment.

5. CONCLUSION

Our approach efficiently provides update traffic management information in a cost-effective manner due to use the black box video data which already exist with each vehicle. Both video and sensor data gathered automatically from the black box. It is straightforward than the traditional method we can also use the mobile network to communicate at the time of emergency. Our approach still needs to apply to the different type of dataset as well as benchmark dataset. For the experiment, we transfer the data from flash memory to computer hard disk drive. In future, we want to make it more efficient vehicle detection with higher accuracy.

ACKNOWLEDGMENTS

This research was supported by Korean MSIT (Ministry of Science and ICT) under the ITRC support program (IITP-2017-2016-0-00312) supervised by the IITP.

REFERENCES

Michael f. Goodchild (2000). GIS and Transportation: Status and Challenges, Geo Informatica 4:2, 127-139.

Jean-ClaudeThill (2000). Geographic information systems for transportation in perspective, Transportation Research Part C: Emerging Technologies Volume 8, Issues 1–6, February–December 2000, Pages 3-12

Liu R., Servières M. and Moreau g (2012). A GIS data updating method with video, In Proceedings of the International Conference on Computer Graphics Theory and Applications (GRAPP-2012), pages 511-514

Yushi Chen, Zhouhan Lin, Xing Zhao, Gang Wang, and Yanfeng Gu,Member (2014). Deep Learning-Based Classification of Hyperspectral Data, IEEE journal of selected topics in applied earth observations and remote sensing, vol. 7, no. 6,

G. Salvi (2014). An Automated Nighttime Vehicle Counting and Detection System for Traffic Surveillance, 2014 International Conference on Computational Science and Computational Intelligence.

Joel Gonc¸alves1, Joao S. V. Gonc¸alves 2, Rosaldo J. F. Rossetti (2014).Smartphone Sensor Platform to Study Traffic Conditions and Assess Driving Performance, IEEE 17th International Conference on Intelligent Transportation Systems (ITSC)

Ridha Soua, Arief Koesdwiady, and Fakhri Karray (2016). Big-Data-Generated Traffic Flow Prediction Using Deep Learning and Dempster-Shafer Theory International Joint Conference on Neural Networks (IJCNN)

Choosumrong, S., & Raghavan, V. (2011). Real-time cost updates based on current traffic condition for optimal routing planning. In Proceedings of the 32nd Asian Conference on Remote Sensing (ACRS 2011), Taipei, Taiwan (pp. 3-7).

Sananmongkhonchai, S., Tangamchit, P., & Pongpaibool, P. (2008). Road traffic estimation from multiple GPS data using incremental weighted update. In ITS Telecommunications, 2008. ITST 2008. 8th International Conference on (pp. 62-66). IEEE.

Trajcevski, G., Wolfson, O., Xu, B., & Nelson, P. (2002). Real-time traffic updates in moving objects databases. In Database and Expert Systems Applications, 2002. Proceedings. 13th International Workshop on (pp. 698-702). IEEE.

Liu, R., Servières, M., & Moreau, G. (2012). A GIS data updating method with video. In GRAPP 2012 IVAPP 2012-Proceedings of the International Conference on Computer Graphics Theory and Applications and International Conference on Information Visualization Theory and Applications (pp. 511-514).

Alkan, M., & Jacobsen, K. (2015). Information content capabilities of very high resolution optical space imagery for updating GIS database. In SPIE Sensing Technology Applications (pp. 94810K-94810K). International Society for Optics and Photonics.

Champion, N., Stamon, G., & Deseilligny, M. P. (2009). Automatic GIS Updating from High Resolution Satellite Images. In *MVA* (pp. 374-377).

Li, C., Liu, Y., Wang, Q., & Tai, Y. (2008). A GIS database updating algorithm based on GPS surveying data. In Computer Science and Software Engineering, IEEE International Conference on (Vol. 2, pp. 834-837).

Weis, M., Müller, S., Liedtke, C. E., & Pahl, M. (2005). A framework for GIS and imagery data fusion in support of cartographic updating. Information Fusion, *6*(4), 311-317.

Kadri-Dahmani, H., & Osmani, A. (2003). Updating data in GIS: how to maintain database consistency. In Enterprise information systems IV (pp. 163-169). Kluwer Academic Publishers.

Chung, Y., Song, T. J., & Kim, J. (2017). Injury Severity Analysis in Vehicle-Pedestrian Crashes. In International Conference on Applied Human Factors and Ergonomics (pp. 884-892). Springer, Cham.

Mohamedaslam, C., Najeeb, N. A., & Nisi, K. (2016). A smart vehicle for accident prevention using wireless blackbox and eyeblink sensing technology along with seat belt controlled ignition system. In Green Engineering and Technologies (IC-GET), 2016 Online International Conference on (pp. 1-6). IEEE.

Chung, Y., & Chang, I. (2015). How accurate is accident data in road safety research? An application of vehicle black box data regarding pedestrian-to-taxi accidents in Korea. Accident Analysis & Prevention, *84*, 1-8.

Prasad, M. J., Arundathi, S., Anil, N., & Kariyappa, B. S. (2014). Automobile black box system for accident analysis. In Advances in Electronics, Computers and Communications (ICAECC), 2014 International Conference on (pp. 1-5). IEEE.

Han, I., & Yang, K. S. (2009). Characteristic analysis for cognition of dangerous driving using automobile black boxes. International journal of automotive technology, 10(5), 597-605

Lin, C. C., & Wang, M. S. (2010). An implementation of a vehicular digital video recorder system. In Green Computing and Communications (GreenCom), 2010 IEEE/ACM Int'l Conference on & Int'l Conference on Cyber, Physical and Social Computing (CPSCom) pp. 907-911).

Kang, C., & Heo, S. W. (2017). Intelligent safety information gathering system using a smart blackbox. In Consumer Electronics (ICCE), 2017 IEEE International Conference on (pp. 229-230). IEEE.

Moon, H. M., Kim, K. H., Lee, M., Kim, P., & Pan, S. B. (2015). A study on the context aware system for intelligent automobile black box. In Control and Decision Conference (CCDC), 2015 27th Chinese (pp. 2536-2540). IEEE.

Baker, W. E., & Sinkula, J. M. (2002). Market orientation, learning orientation and product innovation: delving into the organization's black box. Journal of market-focused management, 5(1), 5-23.

S. Y. Yi, J. H. Ryu, and C. G. Lee (2010). Development of embedded lane detection image processing algorithm for car black box, Journal of the Korea Academia Industrial Cooperation Society, Vol. 11, No. 8, 2942-2950.

Karen Simonyan, Andrea Vedaldi, Andrew Zisserman(2013). Deep Inside Convolutional Networks: Visualising Image Classification Models and Saliency Maps, arXiv: 1312.6034

Ross Girshick (2015). Fast R-CNN - ICCV 2015

Andrea Vedaldi, Karel Lenc (2015). MatConvNet: Convolutional Neural Networks for MATLAB,Proceedings of the 23rd ACM international conference on Multimedia.

Proc. 1st International Conference on
Machine Learning and Data Engineering (iCMLDE2017)
20-22 Nov 2017, Sydney, Australia

Using Genetic Algorithm with Triple Crossover to Solve Travelling Salesman Problem

Shamima Akter[1],Mohammad Sanaullah Chowdhury[2],Subrina Akter[3],Lutfun Nahar[4] and Md. Wahid Murad[5]

[1,2]Department of CSE, University of Chittagong, Bangladesh
[3,4]Department of CSE, International Islamic University Chittagong, Bangladesh
[5]Senior Software Engineer, DataSoft Systems Bangladesh Limited

Corresponding author's E-mail: sana1691@gmail.com

Abstract

TSP is a popular and demanding NP hard problem. Many researchers get interest into it. There are a numerous methods such as SCX, ERX, and GNX etc. for solving TSP with GA. In our paper we proposed a new solution for Traveling Salesman Problem (TSP) using genetic algorithm. A triple crossover technique is applied for finding best & optimal solution of this problem. Triple Crossover Operator (TCO) separate the parent's strings into three substrings comparing cost and derive new partial Offspring containing duplicate nodes then replacing the missing nodes with duplicate nodes allows the system to generate high performance chromosomes. This solution is compared with different well performing Crossover technique. Our Experimental result shows that, due to the well crossover technique has improved performance. Moreover the complexity of this algorithm is negligible.

Keywords: TSP; GA; TCO; SCX; ERX; GNX; substring

1. INTRODUCTION

Travelling Salesman Problem (TSP) is very popular and challenging problem in computer science and computational research. This problem can be stated as, travel all the nodes in a city exactly once and then back to the starting nodes with minimum cost. This is a NP hard problem and cannot be solved exactly in polynomial time. The TSP applied in different situations like automatic drilling of printed circuit boards and threading of scan cells in a testable VLSI circuit as suggested by ravikumar (1992), X-ray crystallography as suggested by zakir (2010), etc.

Many different methods of optimization have been used to solve TSP, like Tabu search, Simulated Annealing, Particle swarm, Hill climbing and Genetic algorithm as suggested by davis (1985) , Po.n and Carter(1995) etc. .Jun Li (1992) proposed an algorithm for solving TSP problem with exclusive and ordinary cities. The algorithm amended two-chromosome format and planned three pairs of crossover and mutation operators. Aybers et al (2009) suggested a method using Euclidean TSP. This is a NP – hard problem associated with resolving the shortest path through a known set of nodes in n-dimensional Euclidean space.

Another existing parallel GA performance to solve the TSP is done by davis (1985). In this paper order crossover and 2-opt mutation method is done. This paper calculates the different elements among populations. The crossover and mutation parts belong to inner parallelization. The differentiation in the crossover part is to create one new Offspring with the combination of two parents. In our paper we choose Genetic Algorithm to solve TSP to find an optimum solution by proposing a new crossover technique.

GAs stores a population of chromosomes, each of which is a candidate solution for its corresponding problem. In each generation (iteration) of the heuristic, several operations are performed on the chromosomes to improve the overall fitness (i.e., cost) of the population.

The genetic algorithm process consists of the following:

1) Encoding: An appropriate encoding is choose for the solution to our problem so that each possible solution has distinct encoding and the some form of a string is called encoding.
2) Evaluation: Next step is to select the initial population, usually at random though different techniques using heuristics have also been proposed. By using the fitness function each individual in the population is then computed. The nearest fittest individual is choosing for the next process.
3) Crossover: The fittest population is used to find the individual's probability of crossover. Crossover is where the two individuals are recombined to create new offspring which are copied into the new population space. In our paper we use a new crossover operation to make new population.
4) Mutation: Next mutation occurs .Mutation is a genetic operator used to maintain genetic diversity from one generation of a population of chromosomes to the next. The character in the corresponding position of the string is changed.
5) Decoding: In this stage the process is complete. Finally a new generation has been formed and the process is repeated until our optimize criteria has been met. At this point the individual which is closest to the optimum is decoded.

2. METHODOLOGY

We Genetic algorithms (GAs) are based essentially on mimicking the survival of the fittest among the species generated by random changes in the gene-structure of the chromosomes in the evolutionary biology as suggested by goldverg (1989). In order to solve any real life problem by GA, two main requirements are to be satisfied:
(a) A string can represent a solution of the solution space
(b) An objective function and hence a fitness function which measures the goodness of a solution can be defined as suggested by zakir (2010).

2.1. Proposed Genetic Algorithm for Solving TSP

1) At first initial populations of individual strings of nodes are created randomly for the selected TSP problem. Create a cost matrix of the path between every two cities.
2) By using fitness function $F(x) = 1/f(x)$ we assign fittest value to each chromosome.
3) New offspring population is created from two parent chromosomes using proposed crossover operator TCO.
4) Mutation is applied if required.
5) Repeat steps 3 and 4 until we get an optimal solution to the problem.

Below we illustrate our method with an example.

A) Chromosome Design
To solve TSP using GA, at first we need to represent population. In population each chromosome is represented as a sequence of nodes, where each node represent a city and after travelling all nodes once; need to reach the starting node. If the TSP has 6 cities then the chromosome length will be 5.The cost of travelling between two cities is represented as a cost matrix which we use in our paper. Figure 2 shows a sample chromosome for a problem with 6 cities:

B) Fitness Function
The TSP problem is solved by the GA which is used for maximization problem. We need a fitness function that define by, $F(x) = 1/f(x)$

Where, f(x) is the objective function which computes the total cost of a tour represented by a string. Chromosomes are selected with a probability included with their fitness value and then copied into next generation. This process is done in selection process. Highly fittest chromosomes can prevent losing the best found solution.

C) Crossover operator
New solution space is made by creating new offspring's from earlier ones. Then we applied our crossover operator in new solution space. At first, two parents are randomly selected. Then we search for two crossover point, one is from starting and another is from ending side. Then we compare and select the fittest portions from parents. After this, if any duplicate node is found in newly generated chromosome, replace the node with untouched nodes again by comparing the cost with previous adjacent node. The algorithm for this newly crossover technique is as follows:

Step 1: Randomly select two parents P1 and P2 and measure their cost. The representation is illustrates in figure 3.

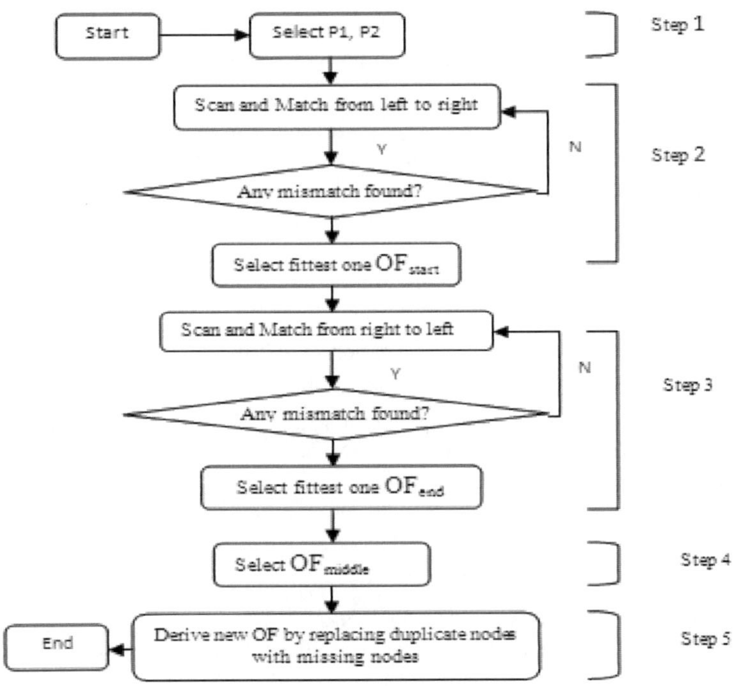

Figure 1. Flowchart of our proposed method

1 4 3 6 2 5

Figure 2. Chromosome representation

P1	4	3	5	7	6	1	2	Cost=282
P2	4	5	6	1	3	7	2	Cost=329

Figure 3. Parent chromosome representation

Step 2: Start from the first node of both parents and scan from left to right to check whether there is any different sequence or not. If found then stop scanning & compare the cost of last two scanned nodes of both parents and select the fittest one OF (start).
A mismatch is found on 2nd position, so we compare the cost of 4-3(11) and 4-5(59), and select 4-3 as it gives the minimum cost. OF start: 4-3

OF 4 3 ○ ○ ○ ○ ○ Cost=11

Figure 4. Partial chromosome representation of offspring

Step 3: Start from the last node of both parents and repeat step 2 & select the fittest one OF (end).
Now, start from the last node we found an mismatch at 2nd last position so we compare between 1-2(75) and 7-2(31) and select the lowest one.OFend:7-2

OF 4 3 ○ ○ ○ 7 2 Cost=42

Figure 5. Partial chromosome representation of offspring

Step 4: Select OF (middle) by comparing the cost of the parent's no scanned middle substrings.
Now we compare middle nodes which are not scanned 5-7-6(cost-132) from P1 and 6-1-5(cost-71) from P2 and select the minimum one 6-1-5 with 71.
OF_{middle}: 6-1-5.

OF 4 3 6 1 5 7 2 Cost=230

Figure 6. Final chromosome representation of offspring

Step 5: List out the missing nodes (Mi) from the newly derived immature OF. Replace the duplicate nodes (Di) with the missing nodes (Mi) from left to right by considering the minimum cost between their previous adjacent node (Ai).
If (Ai-Di) < (Ai-Mi)
Select Di
Else
Select Mi
Step 6: Finally Derived new Offspring which requires the lowest cost.
To explicit the method, a problem with 7 cities is considered. Cost matrix is included in table I.

Table 1. Portion of a cost matrix

Node	1	2	3	4	5	6	7
1	100	75	99	9	35	63	8
2	51	100	86	46	88	29	20
3	50	5	100	16	28	35	28
4	20	45	11	100	59	53	49
5	86	63	33	65	100	76	72
6	36	53	89	31	21	100	52
7	58	31	43	67	52	60	100

3. COMPUTATIONAL RESULT AND DISCUSSION

In this age of modern science we concern with time rather than memory. Our proposed method takes a number of calculations but less iteration which utilizes memory but saves time. The previous methods were very good but our proposed method gives optimal solution in less time than the previous methods.

For comparing the efficiency of the different crossover operators, genetic algorithms using SCX, ERX, GNX and TCO have been encoded in Visual C++ on a Pentium 4 personal computer with speed 5 GHz and 1GB RAM under MS Windows 7, and for some TSPLIB instances. Initial population is generated randomly. The following common parameters are selected for the algorithms: population size is 300, probability of crossover is 1.0 (i.e., 100%), probability of mutation is 0.01 (i.e., 1%), and maximum of 15,000 generations as the terminating condition. The experiments were performed 8 times for each instance. The solution quality is measured by the percentage of excess above the optimal solution value reported in TSPLIB website, as given by the formula

$$Excess(\%) = \frac{(Solution\ Value\ -\ Optimal\ Solution)}{Optimal\ Solution\ ValueValue} \times 100\% \quad (1)$$

We report percentage of excess of best solution value and average solution value over the optimal solution value of 10 runs. The table also reports the average time of convergence (in second) by the algorithms.

Table 2. Summary of the results of different crossover operators for asymmetric TSPLIB instances

Instance	n	Opt. Sol.	ERX Avg Time	GNX Avg Time	SCX Avg Time	TCO Avg Time
br17	17	39	2.10	0.26	0.11	0.9
ftv33	34	1286	70.22	7.64	2.25	1.98
ftv35	36	1473	76.39	1.48	9.69	6.92
ftv38	39	1530	160.87	4.03	6.89	7.12
p43	43	5620	213.99	22.33	22.98	20.22
ftv44	45	1613	157.23	18.46	19.22	15.34
ftv47	48	1776	200.70	42.67	25.99	18.10
ry48p	48	14422	185.59	39.33	25.73	27.31
ft53	53	6905	122.75	29.35	36.73	31.11
ftv55	56	1608	328.59	23.74	35.11	35.21
ftv64	65	1839	326.95	91.39	76.56	81.01
ft70	70	38673	561.14	90.13	74.19	73.99
ftv70	71	1950	432.31	135.97	58.69	58.14
kro124p	100	36230	542.57	178.64	142.02	140.01
ftv170	171	2755	526.46	483.21	259.60	248.91

We also measure the performance of different crossover operators for the instances ftv170 and for some symmetric TSPLIB instances (considering 500 generations). For these instances our proposed method perform better than the previous methods.

4. CONCLUSION AND FUTURE WORK

Solving TSP by Genetic Algorithm is purely depending on the way problem is encoded and which crossover and mutation technique is used. Here we proposed a new crossover operator to solve TSP along with other steps of Genetic Algorithm. In comparison with other crossover operator like SCX, ERX as defined by zakir (2010) our proposed technique shows the best result in terms of times. We use a local search technique to select initial population. In future we want to find a search technique to minimize initial population with good quality of chromosomes which can give a better result by minimizing iteration as well as cost and time.

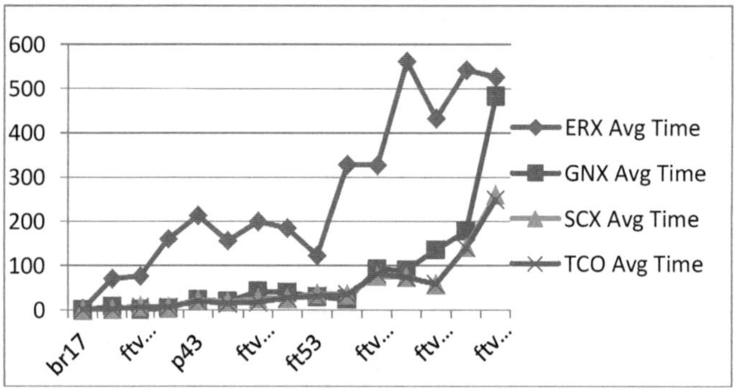

Figure 7. Performance analysis of different crossover operators for asymmetric TSPLIB instances

5. REFERENCES

Jun L, Qirui S, MengChu Z, Xianzhong D (1992). A New Multiple Traveling Salesman Problem and its Genetic Algorithm-based Solution.

C.P. R (1992), Solving Large-scale Travelling Salesperson Problems on Parallel Machines, Microprocessors and Microsystems 16(3), pp. 149-158.

Zakir H. A (2010). Genetic Algorithm for the TSP using Sequential Constructive Crossover Operator, International Journal of Biometrics & Bioinformatics (IJBB), Volume (3), Issue (6).

Aybars U, Serdar K, Ali C, Muhammed C, Ali A (2009). Genetic algorithm based solution for TSP on a sphere, Mathematical and Computational Applications, Vol. 14, No. 3, pp. 219-228.

L. D (1985). Job-shop scheduling with Genetic Algorithms, Proceedings of an International Conference on Genetic Algorithms and Their Applications, pp. 136-140.

C. M, J. K, G. C and Y. S (2002). An efficient genetic algorithm for the traveling salesman problem with precedence constraints, European Journal of Operational Research 140, pp. 606-617.

N.J. R and P.D. S (1995). Formae and variance of fitness, In D. Whitley and M. Vose (Eds.) Foundations of Genetic Algorithms 3, Morgan Kaufmann, San Mateo, CA, pp. 51-72.

D.E. G (1989). Genetic Algorithms in Search, Optimization, and Machine Learning, Addison Wesley, New York.

Proc. 1st International Conference on Machine Learning and Data Engineering (iCMLDE2017) 20-22 Nov 2017, Sydney, Australia

MPU Based 6050 for 7bot Robot Arm Control Pose Algorithm

Liuqiang Qian[1] , Zhengwei Zhu[2]andCheng Liu[3]
[1]Changzhou university , Changzhou, China
[2]Changzhou university , Changzhou, China
[3]Changzhou university , Changzhou, China

Corresponding author's E-mail: zhuzw@cczu.edu.cn

Abstract

In recent years, domestic and foreign researchers using the robot arm in traditional manufacturing industry, medical treatment, space exploration, education and other fields and made some successful results. The accuracy and stability of the manipulator has been an important topic of many scholars, especially in some specific areas ex) industrial assembly, security, explosion prevention, medical and other fields with special requirements ex) higher precision and movement are required to produce very small deviations. on the mechanical stability arm. This paper introduces the attitude operator sensor MPU6050 with 7BOT multi DOF robot arm manipulator to improve the stability of extremely precise experiments.I fix MPU6050 on 7BOT mechanical arm, MPU6050 mechanical arm posture by clearing the number According to the esp8266 wireless module to transmit to the server, the waveform and generation of 3D position speed and angular speed, including processing and produce the 3D mechanical arm motion model, while the 7BOT attitude data sent to the server, automatic data and waveform than on. If large deviations are corrected in a timely manner, the revised data sent to the 7BOT arm, arm posture timely correction. The experiment proved that can greatly improve the mechanical arm to complete a given task accuracy in initial position.

Keywords: Attitude calculation, MPU 6050, 7BOT Robot Arm,

1. INTRODUCTION

Mpu6050 attitude sensor is used in many kinds of application scenarios. L. Anusha and Y. U. Devi Implementation of gesture based voice in and language translator for dumb people in the MPU6050 collection with the gesture recognition algorithm of precise orbit information back to the deaf, and converted into normal language output in Implementation of IMU s. Ensor for elbow movement measurement of Badminton players Jacob W. N. the A., Wan Zakaria and M. R. B. Md Tomari MPU6050 for arm joints for athlete's motion capture system, can accurately capture the athlete's movement speed, improve the efficiency of data of athlete analysis. In the D Bassily C Georgoulas,, J Gue Ttler, T Linner, T Bock Intuitive and Adaptive Robotic Arm Manipulation using the Leap Motion Controller the mechanical arm through the combination of leap motion, to implement the real-time control of the manipulator by leap motion gesture recognition, greatly reducing the control of the mechanical arm threshold.

In addition MPU6050 can be widely to improve the mechanical arm movement In the Trajectory optimization using time-separating efficiency, strategy with improved PSO on mechanical arms Meng P. Feng the Z., P., Chao, L. Weixing and G. Qi is a combination of attitude sensor and particle swarm optimization algorithm for joint trajectory optimization, the test results show that can greatly improve the accuracy of the manipulator trajectory . The use of mechanical arm is very broad, recently in A. Joly, R. Zheng and K. Nakano A Scaling Method for Real-Time Monitoring of Mechanical Arm Admittance, "a text, the author proposes a manipulator to obtain the parameters of road safety system driving intention to timely modify the driving error based on, this paper proposes a to improve the method of manipulator calibration. In K. Nishikawa, K Furukawa, I. Kawate, T. Miyazaki, T. Nouzawa and T. Tsuji Design of steering wheel characteristics based on human arm mechanical properties in an article by mechanical arm and arm compared to the direction of vehicle design disc system. In the Y. B. Li and Z. L. Jin Kinematic analysis of a Novel 3-DOF hybrid mechanical arm, "a paper mechanical arm motion analysis plays an important role in the application of mechanical arm, according to each different mechanical arm manipulator is established, inverse displacement equation, can fast and improve the operation efficiency and accuracy of the manipulator.

2. RELATED WORK

In general, the manipulator of predetermined initial position and the actual initial position will have a certain deviation, then make some actions according to the demand will have a certain deviation between the actual position and the predetermined position, the main reason is caused by the mechanical structure of the 7bot manipulator, 7bot joint in according to the need for the moving distance is according to the rotation angle number of turns of mechanical arm joint gear to the actual control, so the distance and angle of the actual control of the number of laps the mechanical arm movement will have a certain gap in different types of the same mechanical arm. In general, the manipulator of predetermined initial position and the actual initial position will have a certain deviation, then make some actions according to the demand will have a certain deviation between the actual position and the predetermined position, the main reason is caused by the mechanical structure of the 7bot manipulator, 7bot joint in according to the need for the moving distance is according to the rotation angle number of turns
of mechanical arm joint gear to the actual control, so the distance and angle of the actual control of the number of laps the mechanical arm movement will have a certain gap in different types of the same mechanical arm.

This experiment will be 6 degrees of freedom MPU6050 attitude sensor and 7bot manipulator with combination of trajectory optimization algorithm, algorithm is improved by 7 BOT mechanical arm to complete the established rate and accuracy of sensor fusion. In Chongran Jiang, Qingjun Shi,

Wenping Chen and Binshan Xu Research and design on distributed controllers for mechanical arms of Inertial sensors are used in humanoid robot, and the advantages of using sensors to optimize robot arm trajectories and prevent excessive fluctuations are considered: small size, low cost and so on.

3. MAIN GOAL

This algorithm mainly through the attitude calculation based on sensor to solve the manipulator control system leap motion control problem is inevitable because the desktop manipulator itself the mechanical structure of the angular position error.

Figure 1 is show about 7bot robot arm initial pose problem. When initial pose value to 7bot robot arm, it's Arduino due board programing setting servo motors to initial pose Like {85, 120, 60, 90, 90,90} the value is factitious. It's changlle to setting value to accuracy. So need other approach checking pose estimation, this paper support pose estimation using MPU 6065 make it more better.

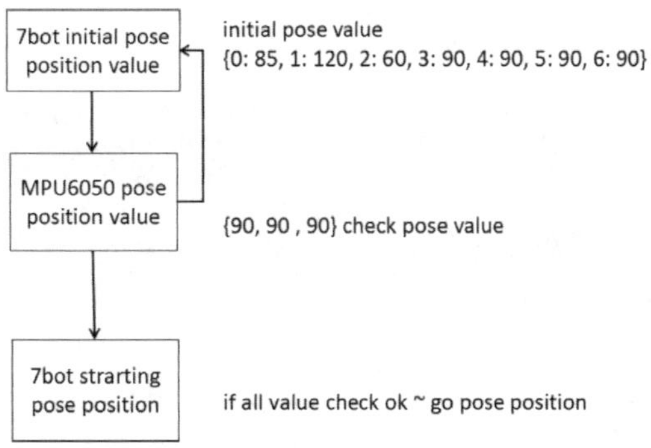

Figure 1 . Flowchart initial position check algorithm

The attitude sensor will produce settlement data drift phenomenon, this study uses Kalman filter and two order filter algorithm of attitude sensor is fixed on the back arm above the data filtering, the actual movement of the 3D model of mechanical arm movement is extremely close to the mechanical arm. The initial value of 7bot value for the initial steering, when the steering gear assembly is difficult to achieve accurate initial value. If the initial position angle of 90 degrees 7bot reality should be the value is set to the value of the steering gear. This value is consistent after the initial pose and initial value. When the need to change the posture change of steering gear the value to control. This benefits cost feedback value will be very fast. When the frequency value does not appear The initial value can only be made. For example, after being used to replace the 7bot actuator to the initial value of angles = {0: 85, 1: 120, 2: 60, 3: 90, 4: 90, 5: 90, 6: 90} 1 steering gear had been biased 5, other position also address actuator is not precise. Because the 7bot value to determine the position of reverse gear can see its driver C language inside filter three ForceFilter.h, MedianFilter.h, PressFilter.hforcefilter for mechanical arm mode of protection, soft mode and normal mode corresponding to each of its function, normal pause change value until the next input value position, soft mode to pause servo value feedback, in order to handle staff after moving this process value preservation and 3-D visualization of this process. Protected mode is reserved for pauses and initial values or current values. Figure 2 is show the six axis sensor, which is in three directions: velocity, angular velocity, and acceleration data. Figure 3 is show MPU 6050 gyro sensor and acceleration of gravity transfrom psoe to x,y,z 3D space, using this value to checking 7bot pose estimation is accuracy problem. Improve 7bot initial pose value accuracy.

```
MPU6050_ID:, 0x68, ACC_X:, 7748, ACC_Y:, -3359, ACC_Z:, -2229 ,GYRO_X:,  -42, GYRO_Y:, -189, GYRO_Z:,  28
MPU6050_ID:, 0x68, ACC_X:, 7747, ACC_Y:, -3357, ACC_Z:, -2228 ,GYRO_X:,  -42, GYRO_Y:, -203, GYRO_Z:,  27
MPU6050_ID:, 0x68, ACC_X:, 7751, ACC_Y:, -3355, ACC_Z:, -2230 ,GYRO_X:,  -39, GYRO_Y:, -192, GYRO_Z:,  28
MPU6050_ID:, 0x68, ACC_X:, 7748, ACC_Y:, -3356, ACC_Z:, -2234 ,GYRO_X:,  -40, GYRO_Y:, -191, GYRO_Z:,  28
MPU6050_ID:, 0x68, ACC_X:, 7753, ACC_Y:, -3365, ACC_Z:, -2238 ,GYRO_X:,  -37, GYRO_Y:, -191, GYRO_Z:,  27
MPU6050_ID:, 0x68, ACC_X:, 7750, ACC_Y:, -3357, ACC_Z:, -2236 ,GYRO_X:,  -40, GYRO_Y:, -188, GYRO_Z:,  29
MPU6050_ID:, 0x68, ACC_X:, 7752, ACC_Y:, -3356, ACC_Z:, -2227 ,GYRO_X:,  -41, GYRO_Y:, -189, GYRO_Z:,  29
MPU6050_ID:, 0x68, ACC_X:, 7746, ACC_Y:, -3356, ACC_Z:, -2249 ,GYRO_X:,  -40, GYRO_Y:, -190, GYRO_Z:,  28
MPU6050_ID:, 0x68, ACC_X:, 7748, ACC_Y:, -3352, ACC_Z:, -2251 ,GYRO_X:,  -42, GYRO_Y:, -191, GYRO_Z:,  29
MPU6050_ID:, 0x68, ACC_X:, 7767, ACC_Y:, -3310, ACC_Z:, -2281 ,GYRO_X:,  -44, GYRO_Y:, -208, GYRO_Z:,   0
MPU6050_ID:, 0x68, ACC_X:, 7764, ACC_Y:, -3313, ACC_Z:, -2280 ,GYRO_X:,  -42, GYRO_Y:, -190, GYRO_Z:,  28
MPU6050_ID:, 0x68, ACC_X:, 7763, ACC_Y:, -3308, ACC_Z:, -2284 ,GYRO_X:,  -43, GYRO_Y:, -192, GYRO_Z:,  24
MPU6050_ID:, 0x68, ACC_X:, 7773, ACC_Y:, -3297, ACC_Z:, -2277 ,GYRO_X:,  -41, GYRO_Y:, -188, GYRO_Z:,  30
MPU6050_ID:, 0x68, ACC_X:, 7769, ACC_Y:, -3288, ACC_Z:, -2277 ,GYRO_X:,  -41, GYRO_Y:, -188, GYRO_Z:,  29
MPU6050_ID:, 0x68, ACC_X:, 7773, ACC_Y:, -3300, ACC_Z:, -2291 ,GYRO_X:,  -41, GYRO_Y:, -189, GYRO_Z:,  29
MPU6050_ID:, 0x68, ACC_X:, 7769, ACC_Y:, -3299, ACC_Z:, -2287 ,GYRO_X:,  -41, GYRO_Y:, -188, GYRO_Z:,  29
MPU6050_ID:, 0x68, ACC_X:, 7768, ACC_Y:, -3291, ACC_Z:, -2301 ,GYRO_X:,  -41, GYRO_Y:, -189, GYRO_Z:,  30
MPU6050_ID:, 0x68, ACC_X:, 7766, ACC_Y:, -3299, ACC_Z:, -2291 ,GYRO_X:,  -42, GYRO_Y:, -189, GYRO_Z:,  31
MPU6050_ID:, 0x68, ACC_X:, 7764, ACC_Y:, -3306, ACC_Z:, -2286 ,GYRO_X:,  -41, GYRO_Y:, -187, GYRO_Z:,  30
MPU6050_ID:, 0x68, ACC_X:, 7768, ACC_Y:, -3297, ACC_Z:, -2281 ,GYRO_X:,  -41, GYRO_Y:, -189, GYRO_Z:,  30
MPU6050_ID:, 0x68, ACC_X:, 7771, ACC_Y:, -3301, ACC_Z:, -2286 ,GYRO_X:,  -40, GYRO_Y:, -189, GYRO_Z:,  31
MPU6050_ID:, 0x68, ACC_X:, 7768, ACC_Y:, -3301, ACC_Z:, -2289 ,GYRO_X:,  -41, GYRO_Y:, -191, GYRO_Z:,  28
MPU6050_ID:, 0x68, ACC_X:, 7763, ACC_Y:, -3293, ACC_Z:, -2287 ,GYRO_X:,  -42, GYRO_Y:, -190, GYRO_Z:,  28
```

Figure 2. MPU 6050 Six axis sensor data

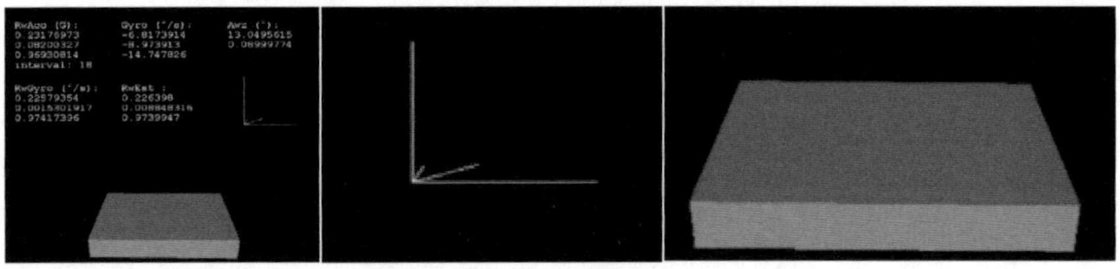

Figure 3. MPU 6050 gyro sensor and acceleration of gravity

4. CONCLUSION

The experiment proved that the sensor can be modified through the gesture manipulator of the actual position and position of the original gap can be corrected because of 7bot desktop manipulator position error is inevitable because of the mechanical structure itself, although in the actual experiment because processing programs such as delay reasons, the mechanical arm position correction will produce minimal delay. But has a significant effect on improving the accuracy and efficiency of the manipulator.

5. ACKNOWLEDGMENTS

This work is supported by the General Programs of China National Science Foundation under grant No. 61772090.

6. REFERENCES

Anusha L, Devi Y(2016). Implementation of gesture based voice and language translator for dumb people, 2016 International Conference on Communication and Electronics Systems(ICCES) ,4, 1-4.

Jacob A, Wan W, Tomari M (2016).Implementation of IMU sensor for elbow movement measurement of Badminton players, 2016 2nd IEEE International Symposium on Robotics and Manufacturing Automation (ROMA),4,1-6.

Meng Z, Feng P, Chao P, Wei X, Qi G.(2017). Trajectory optimization using time-separating strategy with improved PSO on mechanical arms, 2017 36th Chinese Control Conference (CCC),4,1-6

Jiang C, Shi Q, Chen W, Xu B(2010).Research and design on distributed controllers for mechanical arms of humanoid robot,2010 2nd International Asia Conference on Informatics in Control, Automation and Robotics (CAR 2010),3,88-91

Zhang D, Wei B(2016).Design of a joint control system for serial mechanical arms based on PID and MRAC control,2016 Asia-Pacific Conference on Intelligent Robot Systems (ACIRS),10,91-96.

Bassily D, Georgoulas C, Guettler J (2014).Intuitive and Adaptive Robotic Arm Manipulation using the Leap Motion Controller.*ISR/Robotik* 2014,41st International Symposium on Robotics, Munich, Germany,10,1-7.

Joly A, Zheng R, Nakano K(2015).A Scaling Method for Real-Time Monitoring of Mechanical Arm Admittance, 2015 IEEE International Conference on Systems, Man, and Cybernetics,5,1551-1556.

Nishikawa K, Furukawa K, Tsuji T.(2014).Design of steering wheel characteristics based on human arm mechanical properties,2014 IEEE/SICE International Symposium on System Integration,10,508-513.

Zhang J, Fang J (2008).Research of Mechanical Arm Control Based on Data Glove,2008 International Symposium on Information Science and Engineering,10,188-191.

Chao Y, Zhou Y, Xu Y(2013).Mobile robotic control system based on master-slave teleoperate mechanical arm,2013 25th Chinese Control and Decision Conference (CCDC),10,3195-3198.

Li Y, Jin Z(2008).Kinematic analysis of a Novel 3-DOF hybrid mechanical arm,2008 IEEE International Conference on Industrial Technology,10, 1-5.

Identification of Susceptible Websites from Code Injection Attach

Hussein Alnabulsi, Rafiqul Islam

School of Computing and Mathematics, Charles Sturt University, Albury,2640, Australia
School of Computing and Mathematics, Charles Sturt University, Albury,2640, Australia
{halnabulsi,mislam}@csu.edu.au

Abstract

The paper presents a framework methodology for identifying the real website is vulnerable to code injection attack, our proposed methodology gives a solution for thousands of websites to identify their vulnerability against code injection attacks, so it can help the administrator of the website by checking his own website and knows if it is vulnerable to code injection attacks. There are not too many researchers about this subject because most of the cyber securityresearchers are about detecting and protecting against the malware, the good thing in this research is that it can provide a self-checking protection for every website, so the administrator can know if his website needs to get more protection support against malware or if the anti-malware that the administrator is using for protection against code injection attack is enough.
The framework that presented in this paper can do a code injection attacks such as SQL injection and XSS attacks by itself to check whether the website is vulnerable to code injection attack. The checking methodology is contained all ways that the attacker can do to success his code injection attack, so it gives a precision checking for the website whether is vulnerable against code injection attack or no. By this way, we can protect many of websites against hackers that using code injection attack to make damage such as stealing or deleting or altering important data from website systems.

Keywords: code injection attack, XSS attack, SQL injection attack;

1. Introduction

Many organizations and companies have a problem to secure their web application and the database that connected with it, because of the hackers still can do many webs hacking through code injection attacks such as SQL injection attacks and XSS attacks, we build our framework to help organizations and companies to check if their web applications have any vulnerability that let hackers to exploit it to do their malicious attacks such as stealing the dataset of web application or make any sort of damage to the web application.
The basic idea of the paper is to build a framework that injects a code to the web site that we want to check wither is vulnerable or no, We need to specify what the meta-characters or metadata that we can use as an injection code to the website to check its vulnerability against SQL injection attacks and XSS attacks as it shows in table 1.

Table 1. SQL injection and XSS injection attacks Methods

	SQL code injection	XSS code injection
1.	'="or'	<script>alert('XSS attack')</script>
2.	OR 1=1–	<IMG SRC="javascript:alert('XSS');">
3.	OR '0' = '0'	<script>alert(document.cookie)</script>
4.	') OR ('0' = '0'	%22sCrIpt%2Balert(%27XSS%27)%2B/sCrIpt %22
5.	UNION	<Img SRC=javascript:alert(String.fromCharCode(88,83,83))>
6.	'	<Img SRC=# Onmouseover="alert('xxs attack')">
7.	SHUTDOWN	<SCrIPt>Alert(String.fromCharCode(88,83,83))</S CrIPt>

So to know whether the website is vulnerable to code injection attack or not, the framework injects a meta-code to the website, if the injection is success, it will show that there is some problem that happensin the website as it shown in figure 1, we can know if the website is vulnerable to code injection attack or not.

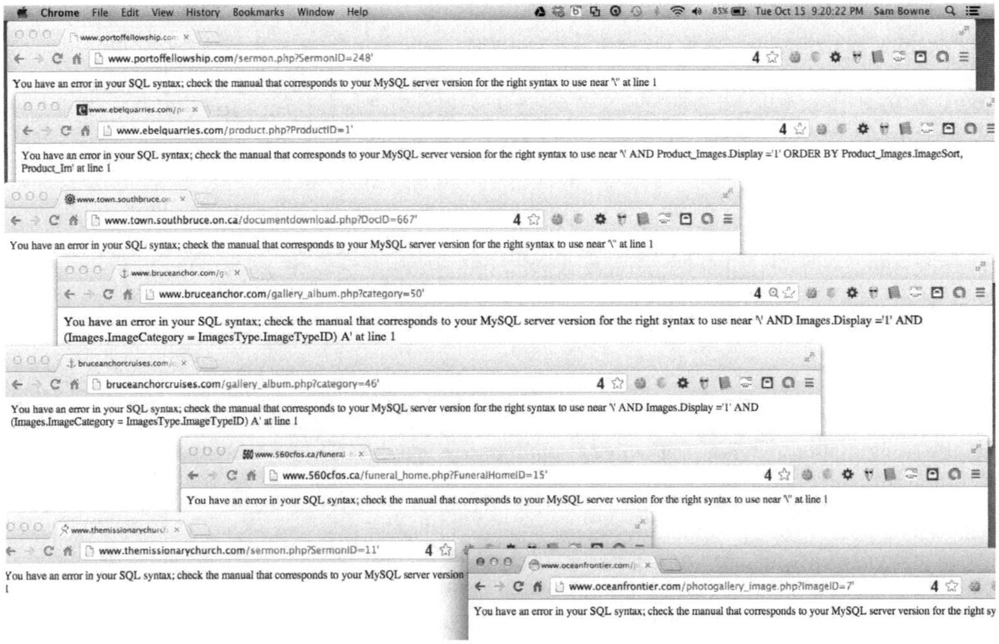

Fig. 1. Some websites that have a vulnerable against SQL injection attacks [11]

In figure 1 the attacker injected a meta-code (') to the web page that connected to the database of the web application to check whether is vulnerable to SQL injection attack, and the results show that there are errors in the SQL syntax and MySQL server, so it means that the websites are vulnerable to SQL injection attacks. We can know that the page of the web application is connected to a database by the URL link, if the URL link of the web application end up with a question mark sign (?) then an equal sign (=) then a number, it means that the web page is connected to the database of the web application.

2. Problem Statement

There are big challenges that the user is totally unaware about the CIA's into web browsers. Lots of companies and websites administrators are suffering from the fact that websites administrators do not know if the company website is vulnerable from CIA or not. Consequently, when a company wants to build a website, the web administrator does not know which protection program the company should use, as it depends on how much the company can afford to buy and use this protection system, or if it is easy enough for the web administrator to use the protection program that the company wants to use as its protection system. So the web administrator uses a protection system according to the cost of it or if the web administrator knows how to treat the protection system. But sometimes the web administrator does not know if the protection system that the company will use is really beneficial for the company to protect against CIA's [12]. So the problem is that a company or a system administrator are not sure if the protection system that they used to protect the website of the company can fully protect the website from CIA's. There are a lot of researchers talking about detection and protection from hacking attacks, and a lot of protection systems can provide some kind of protection against hacking attacks, however, there are no researchers looking into checking if the website is still vulnerable against CIA or not. This issue is important because users are facing this problem of needing an alert system in real life to take preventing measure against CIA's [12].

3. Literature Review

Fonsecaet al. [4] proposed The paper "Mapping Software Faults with Web Security Vulnerabilities", authors classified 655 SQL injection and XSS that they found them in 6 websites applications. Authors conclusion shows that a group of twelve faults in a generic software framework is responsible to all of the security troubles against code injection attacks such as SQL injection and Cross-site scripting (XSS). Authors detected MFC extended which is a part of a Missing Function Call fault, which is responsible of 76% for all the security analyzed problems of code injection attacks. By comparing the distribution of the Missing Function Call fault of the results with researches of software faults authors found considerable differences between them. The detailed analysis of the location or condition of each fault was observed and presented, so it supports the future works for definition the realistic of fault types which effect on detection the vulnerability in a web pages. In the future work, authors want to study the exploit code that an attackers used to Penetrate web pages, the results can help to create an attack injector to Penetrate web pages.

Stott et al. [7] proposed the paper "NFTAPE: a framework for assessing dependability in distributed systems with lightweight fault injectors", In this paper authors presented NFTAPE, that authors used it as an experiments to conduct fault injection automatically, they used it to help them solving these problems monitors, fault injectors, target specific, trigger, driver based, hardware based, debugger based, simulation based, and performance-fault. Authors used fault injection to accelerate the rate of errors happen in the computer system, for analyzing the dependability of the system, by presenting the weakness and evaluating the coverage of fault injection technique. Different tools of fault injection techniques have been presented. These injectors consist of simulated fault Injectors, physical fault injectors, and software implemented fault injectors (SWIFI). SWIFI is an inexpensive, easy technique to develop and runs in program. The architecture component helps to make it easy for using NFTAPE to many of operating system programs such as Windows, Solaris, Lynx, and Linux. Authors used NFTAPE to do experiments on fault injection technique. They presented 2 experiment examples of fault injection technique. The first experiment example authors inject bit errors into the physical layer of a LAN link by using a hardware fault injector. The second experiment example is framework of the real space imaging by using a fault injector of debugger-based. NFTAPE is flexible with an ability to execute simulated-based fault injector, SWIFI, or hardware-based fault injector. the contribution of the authors is the library of reusable components that built for NFTAPE.

Neveset al. [3] proposed the paper " Using Attack Injection to Discover New Vulnerabilities", in this paper authors present a tool called AJECT which is for discovering the security vulnerabilities on web servers/networks applications. AJECT simulates the behavior of a malicious attack by injecting many types of code injection attacks into the computer server target. AJECT produces malicious attacks automatically to perform the malicious attacks to the target computer server, while the application running, it monitors the framework to collect many sorts of information to determine whether the server executed incorrectly (vulnerability exists) or not. For evaluation, authors conducted several experiments with Internet Message Access Protocol (IMAP), which is a protocol used to access and preview electronic mail on a network of the computer server. The experiments show that AJECT utilized to locate distinct types of vulnerabilities such as format strings, information disclosure bugs, and buffer overflows). So AJECT can find many sorts of code injection vulnerabilities, such as the previously mentioned vulnerability.

Fonseca et al.[6] proposed the paper "Testing and Comparing Web Vulnerability Scanning Tools for SQL Injection and XSS Attacks", in this paper authors evaluate web pages vulnerability using commercial scanners, they injected a software faults in the web pages. Authors compare the efficiency of the different commercial scanners tools in the detection of the vulnerabilities. By analyzing the false positives and the detected vulnerabilities, authors evaluate and compare 3 of the commercial web page vulnerabilities scanners to know which one is better in detection of the code injection vulnerabilities in the web pages. the results show that the scanners get many different results but all of them produce a high false negative and false positives percentage values (false positives is in between 20% to 77%). In the future work authors want to apply and evaluate the same methodology to anther web pages that have not been used in this approach methodology, then compare many other approaches methodologies to get a bigger experiments results and to study which sort of programming code that leads to code injection vulnerabilities so it can support the prevention of code injection vulnerabilities.

Bau, et al. [1] proposed the paper "State of the Art: Automated Black-Box Web Application Vulnerability Testing", authors studied the vulnerabilities of black-box web application scanners to detect vulnerabilities of XSS and SQL injection, Cross-Site Request Forgery, Cross-Channel Scripting, Information Disclosure, and Malware. Authors goal is to identify the strengths and the limitations of different tools and to report test results of detect vulnerabilities using different tools. The eight commercial scanners tools that incorporated in authors study are (NeXpose 4.8.0, McAfee-SECURE, Rational AppScan 7.9, WVS 6.5, HailStorm-Pro 6.0, QA-Edition 7.0.0, QualysGuard-PCI, Web-Inspect 8.0), the study includes scanners tools of many of the companies in the web security. The prices of the scanners tools starts from (100's to 10,000's) of dollars, which gives a many options for users and customers to select which scanner tool is more appropriate for them. The scanners tools are poorly in detecting "stored" vulnerabilities, such as the stored XSS detection rate was 15%, scanners tools could not detect the vulnerabilities of the second method of SQL injection. 7 scanners tools detected the vulnerabilities of the first method of SQL injection (which is around 14%). Just one scanner tool has been able to get a high rate of detection the vulnerabilities of the first-order SQL injection (which is around 40%). Authors testbed has a server running Apache server 2.2.3, MySQL 5.0.45, PHP 5.1.6, and the PhpMyAdmin installed on a Linux 2.6.18 was running on the server with around 50 URLs, 3000 lines of code.

Bhojak, et al. [8] proposed the paper "Automated Web Application Vulnerability Detection With Penetration Testing", in this paper authors proposed a new approach to find a vulnerability in code injection attacks that include a black-box on some existing different web applications such as: Mutillidae, WebGoat, Swapp, dvwa. Authors performed tests to scan for all vulnerabilities using the commercial web vulnerability tools scanners, then they compare their open source scanner with the commercial web vulnerability tools scanners, and scan some web applications to get the results for different web vulnerability as presented in table 2.

Table 2. Results of comparison between commercial web vulnerability scanners and Authors Scanner

		Commercial Scanner 1	Commercial Scanner 2	Authors Scanner
Site tested	Vulnerabilities	Number of Vulnerability Detected		
web applications Site 1	Cross Site Scripting Attacks (XSS)	5	7	8
	SQL Injection Attacks	8	7	7
	Directory Listing	11	9	12
	Broken Authentication using SQL Injection	1	1	1
	Auto-complete enabled on input fields Enabled	0	0	0
web applications Site 2	Cross Site Scripting Attacks(XSS)	6	5	8
	SQL Injection Attacks	3	5	6
	Directory Listing	12	9	13
	Broken Authentication using SQL Injection Attacks	0	0	0
	Auto-complete enabled on input fields Enabled	1	0	2

According to table 2, it shows that the authors tool scanner in web applications Site 1 gives a better result in detection of vulnerabilities such as SQL Injection Attacks, Directory Listing comparing with Commercial Scanner 1,2, but the Commercial Scanner 1 gives a better result in detection Cross-Site Scripting Attacks (XSS) comparing with Authors Scanner. In web applications Site 2 the Authors Scanner gives a better results in detection SQL Injection attacks, Directory Listing, XSS attacks, Auto-complete enabled on input fields Enabled, while it gives the same result in detection Broken Authentication using SQL Injection Attacks for both of Commercial Scanner 1,2 and the Authors Scanner.

4. Methodology

The idea in our framework is different than other researchers works, our idea is to depend on building a framework that can do a code injection attacks to the web applications, so in this way, our framework can check wither the web application has a vulnerability against code injection attacks or not.

The methodology: In our program at first level, we enter the link of URL page that we want to check its vulnerability against SQL injection attacks and XSS attacks, then the program gives all pages link that connected with the URL page that we entered before. Soat this level, we will get all URL pages which connected with the URL page, because we need to get the pages that contain the databases of the URL page, so we can do a checking vulnerability in the next step level.

In the second level, we checking the vulnerability for all URL links by injecting a SQL injection code and XSS code attacks for every URL links, in this level we injected SQL code and XSS code such as: injects[0] = "'="or'", injects[1] = "OR 1=1--", injects[2] = "'OR", injects[3] = "%22%2Balert(%27XSS%27)%2B%22", injects[4] = "OR '0'= '0'", injects[5] = "') OR ('0' = '0'", injects[6] = "'UNION'", injects[7] = "'WHERE'", injects[8] = "'".

In the last level, if the injected is successful, the program will not give an error about the code injection attacks otherwise, the code injection attack does not asuccess, so the website is not vulnerable.

We collect the dataset for checking the code vulnerability from different resources and the dataset is from real-world websites links.

The methodology of this paper is different than other methodologies that we presented before which is depend on checking the vulnerability of the web applications by doing a code injection attacks to the website not as other methodology which checking the commercial vulnerability scanners, or by make a protection framework against code injection attack by checking the source code of the same web page whether is vulnerable to code injection attack or not vulnerable.

5. Experimental Setup

In our experiment, we have collected the dataset from 10 open sources such as Antunes, et al., [13], The latest list of SQL injection vulnerable websites 2016 [14], Vulnerable Sites To (Legally) Practice Your Hacking Skills, [15], Download Biggest SQL Vulnerable Website List For 2017 [16], 150 SQL Vulnerable Websites 2017 List, August 2017 [17], List Of web sites Vulnerable For SQL Injection, August 2017 [18], 20 famous websites vulnerable to cross [19], Xss: xss vulnerable sites collection, August 2017 [20], 10000 fresh sqli vulnerable websites list, August 2017 [21].

We got the benign dataset which is around 100 line of code from the paper "Detecting SQL Injection Attacks Using SNORT IDS"[10].

We have 200 line of data code of the dataset we use for test our program framework, we built the program framework using a C# programming language, and we did the checking for each data line of code manually within the program framework.

Proposed Algorithm:

Step 1:

Get the URL web application that a tester wants to check the vulnerability of it and type it in the text box of the searchbutton.

Step 2:

Get all the URL's pages that connected with the URL web application that a tester entered in the text box of the searchbutton.

Step 3:

From step 2, a tester will get the pages that contain the database of the web application which a tester needs to arrive it to do the code injection attack.

Step 4:

Start checking the URL's pages by doing a code injection attack such as SQL injection attacks and XSS attacks.

Step 5:

A tester inject this code on step 4 such as: injects[0] = "'="or'", injects[1] = "OR 1=1–", injects[2] = "'OR", injects[3] = "%22sCrIpt%2Balert(%27XSS%27)%2B/sCrIpt%22", injects[4] = "OR '0'= '0'", injects[5] = "') OR ('0' = '0'", injects[6] = "'UNION'", injects[7] = "'WHERE'", injects[8] = "'", injects[9] = "''".

Step 6:

if the injected is successful, the program will not give an error about the code injection attacks. Otherwise, the code injection attack does not a success, so the website is not vulnerable.

5.1 Algorithm Evaluation

In this subsection, we evaluate our algorithm from 10 different dataset sources, We have a dataset with 200 lines that we collected from different data sources, the results are:

TP: Attack detected where it was an attack

TN: not attack no alarm no detection

TNR= TN/(TN+FP)

Precision = Positive Predictive Value=TP/(TP+FP)=P

Recall = True Positive Rate = Sensitivity=TP/(TP+FN)=R

True Negative Rate is called Specificity.

False Positive Rate is 1 - Specificity.

Accuracy = (TP+TN)/(TP+TN+FP+FN)

TP=83

FP= 6

FN=0

Precision = positive predictive value=TP/(TP+FP)=83/(83+6)=0.9325

Recall = true positive rate = sensitivity=TP/(TP+FN)=83/(83+0)=1

TNR= TN/(TN+FP)=111/(111+6)=0.9487

Accuracy = (TP+TN)/(TP+TN+FP+FN)=(83+111)/(83+111+6+0)=0.9708.

6. Comparison with other Approaches

1. In the paper of "Training Security Assurance Teams using Vulnerability Injection", authors detecting the vulnerability of code injection by changing the code of web pages.
   The changes on the code of web pages do not affect on the running of the web page, the web will run without any errors in execution [5].
   One of these changes is executed, and it is depending on which code is surrounding the function of code, as it shown in the table 3.

Table 3. Changes on the code and the function of the code

	Function of the Code	The Code	Example
1.	If the function is used in an assignment as the only line of code and the variable is not inside a $_GET, $HTTP_GET_VARS, $_POST, $HTTP_POST_VARS PHP variable array	The whole line of code is removed.	Remove the line "$vuln_var = interval($vuln_var);";
2.	If the function is used in an assignment as the only line of code and the variable is inside a $_GET, $HTTP_GET_VARS, $_POST, $HTTP_POST_VARS PHP variable array	Only the function is removed from the code, leaving the argument intact.	Replace "$vuln_var = interval($_GET['vuln_var']);" with "$vuln_var = $_GET['vuln_var'];";
3.	if the variables is inside, the $_GET, $HTTP_GET_VARS, $_POST, $HTTP_POST_VARS PHP variable array	leave the variables in the code, only the function is removed.	Replace "…"'str1'.intval($vuln_var). 'str2'";" with …"'str1'.$vuln_var.'str2'";".

The authors find that the integer variables ("int" as in C programming language) are a main code injection vulnerability target. It has been collected from subtype A and responsible for 45% of all the vulnerabilities that have been founded by the authors. Authors also check if the "int" variables values are not have a non-integer values (Ex. display, assign). The variables that contain an integer values can be a target to the penetration of code injection attacks, and the penetration attack can be success by adding a meta-code to them, such as:
A- Penetration of SQL Injection attacks , Ex. " or 3=3", " or 'D'='D'".
B- Penetration of XSS attacks "<scrIpT>Alert('alert xss attack 1')</scrIpT>".
C- Adding at the begin or at the end of the variables ["], [( ] ,[ )] , [>], [<] ,['].
Authors used commercial vulnerabilities scanners for web pages to check if the commercial scanners can detect the vulnerabilities and asset the conclusion of results with the security penetration teams. Authors used the HP-WebInspect 7.7, and IBM Watchfire-AppScan 7.0 scanners and named them as S1 and S2, authors make it anonymous because of the commercial license does not allow the publication of evaluation of the commercial scanner.
The Test Results:
The precision rate of Commercial Vulnerability Scanners is 20%
The precision rate of Security Teams after Basic Training Period is 30%
The precision rate of Security Teams after Specific Training Period is 70%
Code Inspection Results:
The precision rate of Security Teams after Basic Training Period is 55.56%.
The precision rate of Security Teams after Vulnerabilities Specific Training Period is 100%.
The limitation is that authors did not build a framework for detection the vulnerability of code injection attacks in the WebPages, they test a Commercial Vulnerability Scanners, and training a Security Teams for detection vulnerability of code injection attacks in the WebPages.

2. In the paper of "Automatic Detection and Correction of Web Application Vulnerabilities using Data Mining to Predict False Positives", Web applications have been the biggest source of problems in the security point of view as some of the reports that indicate the increase of web applications attacks in 2012 is around 33%. The main reason for the increasing of web pages attacks is that lot of the users do not have enough knowledge about how to secure the web pages. The authors present a methodology to protect web pages, the methodology is depending on analyze the source code of the web pages then detecting the code

injection vulnerabilities and modifying the source code in the way of correcting the code injection vulnerabilities [9].
The contribution for the web applications security is by fixing the vulnerabilities, and the programmers can learn from their mistakes by practicing and finding the code injection vulnerabilities.
Authors use Static analysis mechanism with a hybrid method to detect code injection vulnerability. That gives a better effective way to detect the code injection vulnerabilities in source code, but it could be reporting many false positives.
The designing and implementing of WAP (Web Application Protection) is consisting of analyzing the PHP programs. Attacker enter the web page by using an entry points (for example: $ GET) then the attacker Penetrate the vulnerabilities through a database query language such as (MySQL query). Many of the attackers using a benign inputs with meta-characters or meta-data such as (', OR). The web pages is protecting by adding sanitization function between the entry points ($ GET) and the database query language (MySQL query).
Table 4 shown the Sanitization function that have been used to remove the SQL injection vulnerability, which is provided by PHP programming language.

Table 4. Sanitization functions used to fix PHP code for SQLI vulnerabilities.

Vulnerability	Entry points	Sensitive sinks	Sanitization functions
SQL Injection Vulnerability	$Get $Post $Cookie SQLI $ REQUEST HTTP-Get-VARS HTTP-Post-VARS HTTP-Cookie-VARS HTTP-Request-VARS	mysql-Query mysql-unbuffered-Query mysql-db-Query	mysql-real-escape-String
		mysqli-Query mysqli-real-Query mysqli-master-Query mysqli-multi-Query	mysql-real-escape-String
		mysqli-stmt-execute mysqli-execute	mysqli-stmt-bind-Param

To check the result if it is false positive or true positive, authors involved getting the source code of the web page and doing code injection attacks for all vulnerabilities. By correcting and validating the data in the webpage source code, authors can stop the bad effect of meta-characters.
The conclusion of the research paper that contain of the 2 framework tools, nine open source programs and the PHP code of NIST.
The reference of SAMATE Dataset is from [http://samate.nist.gov/SRD/].
WAP's detector detected sixty eight vulnerabilities (twenty two of SQL injection attacks and forty six of XSS attacks), and twenty one of false positives. Pixy detected seventy three vulnerabilities (twenty of SQL injection attacks and fifty three of XSS attacks), and forty one of false positives and five of false negatives, WAP-TA detect five vulnerabilities while the other detectors did not detect them).
The Pixy's accuracy is around 44%, WAP-TA's accuracy is around 69%, and WAP's accuracy is around 92%. The limitation is the high false positive rate which is 20%.

3. In the paper of "Vulnerability & Attack Injection for Web Applications", Authors proposed a code injection approach to detect the vulnerabilities of web pages, they used many programs to applied their idea, that gives a precision methodology to check whether the web application is secure or not [2]. Authors used a case studies depend on the real world to validate the ability of the framework to penetrate web pages, so it validate an Intrusion Detection System and 2 web page code injection vulnerability Scanners for SQL injection attacks.
The Dependency Builder is the first component of the code injection program. Dependency Builder is seeking for the files within the input file, the Dependency Builder is targeting PHP file for the injection process of code injection attack.
In PHP program, programmers can include a general file in to another file, (it can be done using one of this statements: require(), include(), require_once(), include_once()) for reutilization Purposes.
The execution of the main file and the included files is processing by the PHP tool and to search on vulnerabilities, meta-code of code injection attack should be injected.
The second component is the Variable Analyzer. it analyzes all the variables of the SQL queries. The Variable Analyzer collects all PHP variables from the PHP code, then the Variable Analyzer seeks on PHP variables in the SQL query.
Table 5 presents queries that injected sent to the database and execution of SQL Injection. It is consist of the basic of SQL Injection attack strings.

Table 5. Basic attack payload string examples

Attack Payload	Strings Expected result of the attack
‘	The query result is an error
Or 2=2	The query result is the override of the query restrictions
‘ Or ‘b’=’b	The query result is the override of the query restrictions
+connection-id -connection-id	The query result is 0
+2-2	The query result is 0
+67-ASCII('A')	The query result is 0
+51-ASCII(1)	The query result is 0

To evaluate the code injection attack tool for Experimental results authors create 3 groups of experiments. In the first group, they injected meta-code into 3 web pages to validate the performance of the attack. In the second group, they tested an IDS for databases so they can validate the efficiency of the intrusion detection system in the detection of code injection attacks that have been injected by the injector.

In the third group, Authors assess 2 web page vulnerability scanners depending on the ability to detect the code injection vulnerabilities.

In the experiments, authors used LAMP (Linux, Apache, Mysql and PHP) approach. The web server Apache works within the Linux server. Authors used three web applications, the first web application is the TikiWiki that used to build wiki which is a website that let users to contribute to it using adding method or modifying method to its contents).

The second web application is the phpBB. which is a LAMP application, and it is used as an Open Source forum solutions.

The third web application is MyReferences. MyReferences consists of thirteen PHP files and it can be used for management of publications such as the conference, the title, the year of publication.

The result shows that the approach is effective in code injection attack tasks, and the intrusion detection system could detect most of the penetration attacks but the approach gets a high rate of false positives and the code injection vulnerability scanners do have some difficulties in detecting most of the vulnerabilities of code injected attacks.

Authors work is close to our work but the limitation is that they did not do the injection or detection tools in a programming framework, they work on commercial scanners tool and they do a code injection attacks manually.

7. Conclusion

This paper presents a framework method to identify the susceptible websites against code injection attacks. It gives a precision performance compared with other works in the same field as this paper showed before. In this paper, the algorithm framework detects various types of CIA's such as SQL injection attacks, Cross Site Scripting attacks (XSS).

The dataset that we used in this paper has been collected from 10 different resources, we collected 200 lines of code data in the dataset that we use for testing our frame work and applied our framework to them to get the performance of our framework.

The accuracy result of this paper is 0.9708%, with a precious performance of precision rate = 0.9325 and recall rate =1, while the other papers one of them gives an accuracy result was 92.1%.

Our idea is different than other paper work such as [1, 5, 9], which depending on injecting a meta-code by the framework on the web application that we want to check its vulnerability against code injection attacks, such as inject SQL injection meta-characters and XSS meta-characters to check whether the website is vulnerable to code injection attack or not, while the other papers depend on checking the vulnerability of web application by using a commercial applications that have been downloaded from the internet, or by checking the source code of the web application and do some modification on the source code of the web application to prevent the vulnerability against code injection attacks.

References

1. Bau, J., Bursztein, E., Gupta, D., Mitchell, J. (2010). State of the Art: Automated Black-Box Web Application Vulnerability Testing, SP '10 Proceedings of the 2010 IEEE Symposium on Security and Privacy,(pp. 332--345). IEEE.

2. Fonseca, J., Vieira, M., Madeira, H. (2009)Vulnerability & Attack Injection for Web Applications, Proceedings of the 2009 IEEE/IFIP International Conference on Dependable Systems and Networks, IEEE.

3. Neves, N., Antunes, J., Correia, M., Verissimo, P., Neves R. (2006). Using Attack Injection to Discover New Vulnerabilities, IEEE/IFIP International Conference on Dependable Systems and Networks, IEEE.

4. Fonseca, J., Vieira, M. (2008) Mapping Software Faults with Web Security Vulnerabilities", IEEE/IFIP Int. Conference on Dependable Systems and Networks, IEEE.

5. Fonseca, J., Vieira, M., Madeira, H. (2008). Training Security Assurance Teams using Vulnerability Injection, IEEE Pacific Rim Dependable Computing conference, IEEE.

6. Fonseca, J., Vieira, M., Madeira, H. (2008). Testing and Comparing Web Vulnerability Scanning Tools for SQL Injection and XSS Attacks. Proceedings of the 13th Pacific Rim International Symposium on Dependable Computing, (pp.365—372). IEEE.

7. Stott, D.T., Floering, B., Burke, D., Kalbarczpk, Z., Iyer, R.K. (2000). NFTAPE: a framework for assessing dependability in distributed systems with lightweight fault injectors, Computer Performance and Dependability Symp., IEEE.

8. Bhojak, P., Shah, V., Patel, K., Gol, D. (2017).Automated Web Application Vulnerability Detection With Penetration Testing. ICRISET2017 International Conference on Research and Innovations in Science, Engineering &Technology, Vol. 2, (pp.177—187). Kalpa.

9. Medeiros, I., Neves, N., Correia, M. (2014). Automatic Detection and Correction of Web Application Vulnerabilities using Data Mining to Predict False Positives, WWW '14 Proceedings of the 23rd international conference on world wide web, (pp. 63—74). IEEE.

10. Alnabulsi, H., Islam, R., Mamun, Q. (2014). Detecting SQL Injection Attacks Using SNORT IDS. Asia-Pacific World Congress on Computer Science and Engineering Conference, (pp. 1--7). IEEE.

11. Websmart, Inc. and 100,000 Vulnerable Websites, https://samsclass.info/125/ethics/smart-websites.htm, September, 2017.

12. Hussein Anabulsi, A Detection Framework against Code Injection Attacks, Charles Sturt University.

13. Antunes, N., Vieira, M., Madeira, H., "Web Services Vulnerabilities", 2008, http://eden.dei.uc.pt/~mvieir

14. The latest list of SQL injection vulnerable websites 2016, August 2017, http://www.sunnytech7.com/pages/latest-2016-list-of-SQL-vulnerable-websites.html

15. Vulnerable Sites To (Legally) Practice Your Hacking Skills, August 2017, https:// www.checkmarx.com/2015/04/16/15-vulnerable-sites-to-legally-practice-your-hacking-skills/

16. Download Biggest SQL Vulnerable Website List For 2017, August 2017, http://www.arkdots.com/sql-vulnerable-website-list/

17. 150 SQL Vulnerable Websites 2017 List, August 2017, https:// www.scribd.com /document/ 325850327/ 150-SQL-Vulnerable-Websites-2017-List

18. List Of web sites Vulnerable For SQL Injection, August 2017, http://listsql.blogspot.com.au/

19. 20 famous websites vulnerable to cross, August 2017, http://thehackernews.com/2011/09/20-famous-websites-vulnerable-to-cross.html

20. Xss: xss vulnerable sites collection, August 2017, http://cyb3rninjas.blogspot.com.au/2015/03/xss-xss-vulnerable-sites-collection.html

21. 10000 fresh sqli vulnerable websites list, August 2017, https://pastebin.com/ATJE7VdZ

Proc. 1st International Conference on Machine Learning and Data Engineering (iCMLDE2017)
20-22 Nov 2017, Sydney, Australia

iCMLDE GCSTMR

Visual Attention Detection: What is the target person looking at?

Partha Pratim Debnath[1], Md. Golam Rashed[2] and Dipankar Das[3]

[1]Lecturer, Bangladesh Army University of Engineering and Technology, Natore, Bangladesh
[2]Assistant Professor, University of Rajshahi, Rajshahi, Bangladesh
[3]Professor, University of Rajshahi, Rajshahi, Bangladesh
parthapratim.ice10@gmail.com, {golamrashed, dipankar} @ ru.ac.bd

Abstract
In this paper, we have described an approach to detect the visual focus of attention (VFOA) of a person from his/her gaze direction with varying lighting conditions and varying distance from the camera and evaluate their performance. The continuous video of the target person is captured and fed into an expert system for further processing. From frame by frame analysis, the head and eyeball of the target person is detected using vector field of image gradient. If the person changes his/her gaze, the corresponding coordinate of pupil also changes. The frontal view of the person is divided into three regions corresponding to three target objects and the gaze direction are detected based on in which region the coordinate of pupil is located in the eye area. Our technique also combines the both Rapid Eye Movement (REM) and head rotation detection, which provides an efficient tool for VFOA tracking. With a low cost camera, it is very hard to work at night and cloudy days because of lack of proper lighting. So considering these limitations, we have developed a system that can successfully track visual focus of attention of the target person by means of sustained and transient attention or distraction and finally control the attention by deploying an external signaling or alarming system.

Keywords: *Visual focus of attention, eye center localization, gaze detection, sustained attention.*

1. INTRODUCTION

Detection and controlling of human's visual focus of attention (VFOA) is a part of active research for many years [1]. However, the researchers of computer vision and artificial intelligence seek for an accurate method for fixation tracking based on object [2] detection by which human attention level may be realized and diversified attention may be controlled. Basically the term "Fixation" refers to a process of maintaining the visual gaze at a single location. Visual fixation is never perfectly steady, fixational eye movement occurs involuntarily, especially for the case of short-term stimuli/attention. There are many real life situations those need the tracking of VFOA of the target person to avoid the critical circumstances. For example, if we can detect the attention/ distraction of the driver in the vehicle [3], we can develop an automated alarming system to alert the driver and bypass fatal accidents. The results of VFOA detection are also an important tool to generate attention patterns of the particular driver and compare it with the others to select the best ones. To accomplish it, most of the vision researchers used wearable sensor based strategies [4]. However, most of the wearable sensors based strategies have a number of limitations for application in detecting and controlling human's VFOA in real world environments. For example, in the context of public social spaces, people may fix their VFOA at particular location based on their own interests and may not be interested in actively engaging with the technology [5]. Thus, in our proposed technique, we are interested in making use of human's VFOA tracking for a wearable-free solution where human do not need to attach markers to themselves or carry special devices so that we may observe them in an unrestricted manner. Here, in this work, we have used low cost USB camera in front of the target person that captures continuous video frames. Vision researchers formerly used solely head tracking [6] or eyeball tracking [7] [8] techniques independently to meet our objectives.

However, in our approach, we have used head tracking and eyeball-tracking techniques in combination for proper detection of visual focus comprising of transients depends on the tracking of the frequency and time of the head rotation as well as rapid eye movement (REM) realization. The accuracy of our technique is highly sensitive to the choice of proper illumination and right distance from camera-especially for the case of low cost camera deployment. In our present study, we conducted experiments in controlled environment and results show that the proposed approach can yield good performance of detecting visual attention of a target human in general situations that offers transient/sustained attention/ destruction and offers optimum cost efficiency.

2. METHODOLOGY

To detect the visual focus of attention of the target person, we need a camera installed at a suitable position so that image can be captured very easily. These continuous images are feed into our proposed system for frame by frame further processing. The overall process is described as follows-

2.1 Head Pose Detection

The main goal of head pose detection is to track the head from the continuous image whether or not the head is in movement. In our work, we have used the Seeing Machine's faceAPI to detect and track the head pose, h_p of the target person. To detect the head, we have used haar cascade classifier.

2.2 Face Points Extraction by Active Shape Model

Our modeling method works by examining the statistics of the coordinates of the labeled points in the head rectangle. In order to be able to compare equivalent points from different shapes, they must be aligned with respect to a set of axes. We achieve the required alignment by scaling, rotating and translating the shape so that they correspond as closely as possible. In this technique, we aim to minimize a weighted sum of squares of distances between equivalent points on different shapes.

2.3 Iris Center Detection

A multistage approach [9] [10]] has been used to detect the iris. A 3-D head tracker [11] detects the head position and the rectangular area in the image. Then the facial feature points are extracted from the active shape model [12]. These points are used to roughly detect the eye regions from the face. The vector field of image gradient (VFIG) is used to detect the iris center.

The VFIG iris center detection technique is described as follows-

Let I_c be the possible iris center and I_{gi} be the gradient vector in position I_{xi}. If I_{di} is the normalized displacement vector, then it should have some absolute orientation as the gradient I_{gi}. We can determine the optical center I_c* of the iris (darkest position of the eye) by computing the dot products of I_{di} and I_{gi} and finding the global maximum of the dot product over the eye image:

$$I_c* = \text{argmax } I_c \{\frac{1}{N}\sum_{i=1}^{N}(P\ )\}$$

Where, $P = (I_{di}{}^{T}\ I_{gi})^2$ and $I_{di} = (I_{xi} - I_c) / (\|I_{xi} - I_c\|_2)$

$i = 1, 2, \ldots, N$ and the displacement vector I_{di} is scaled to unit length in order to obtain an equal weight for all pixel position in the image.

2.4 VFOA Detection

After the successful detection of the head pose and pupil of the eye, the attention/ distraction of the target person is tracked with the help of head and eye rectangle. A subdivision method proposed in the paper [13], is used here. To achieve the optimum efficiency, we have used the "Unequal Partitioning of eye rectangle leaving the uncovered regions" method to track the movement of the eyeball within the eye area. "Partitioning of the head rectangle" technique is also used to detect the head movement.

Now the speed of the eyeball movement and head movement provides us a clue to detect the short or long term attention/ distraction of the target person. We classify the inattention of the target person in two major criteria-

a) Transient

These short term stimuli occur in the subconscious mind. The target person is not aware of it. It happens unwillingly to provide some rest for the brain.

b) Sustained

This occupies a long time [14]. In our experiments we have set it around 8 seconds. The target person offers this type of inattention willingly.

To detect the transient and sustained attention of the target person efficiently in different environments, two different parameters- right distance from the camera and proper lightening needs to be ensured. It is very important to measure the right distance from the camera. The participant should sit within this range of distances so that the images can be efficiently captured by the camera. Moreover, any deviation from this range of distances can be used as a clue to detect the proper inattention (such as drowsiness). For example, video analysis shows that prior to any road accidents, the driver tends to bow down his head towards the steering of the car. Apparently, the driver comes out of the threshold value of the right distance. To track the threshold value of right distance, we have used the trial and error strategy. The right distance is selected based on the highest efficiency of head and eye center detection. Any deviation from the threshold efficiency provides a clue to inattention.

Proper illumination of eye tracking is also a very important factor of gaze detection. The reason behind this fact is that, to trace the eye pupil, our system tries to find out the darkest point within the face rectangle using the vector field of image gradient. In that case, for the different illumination, the efficiency of head and eye center detection varies. Generally, the efficiency increases with the increase of illumination. But we know that the radius of eyeball decreases with the increase of light intensity after a certain value so that minimum amount of light can enter through the eyeball. So, at the higher illumination level the efficiency of eye center tracking also decreases. The threshold value of proper illumination is selected based on the highest efficiency of head and eye center detection

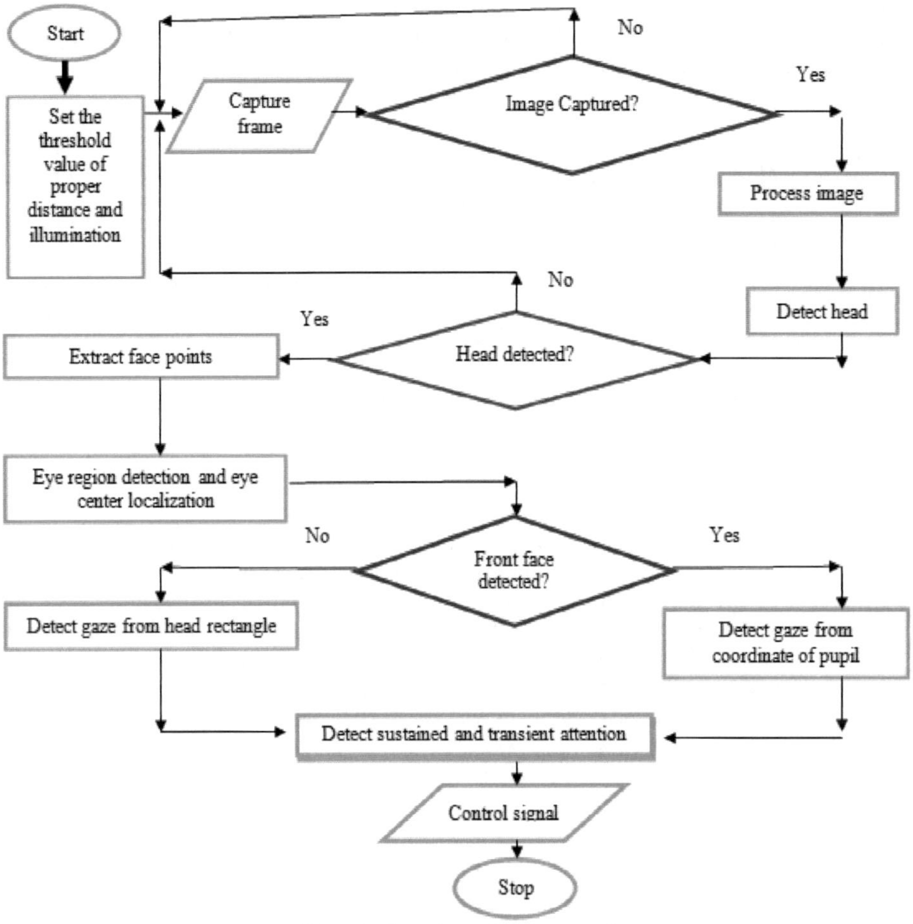

Figure1. Flow chart of the proposed system

3. EXPERIMENTS

3.1 Experimental Setup

We conducted separate experiments applying different techniques under different lighting conditions and distances from camera under a controlled environment. Data were taken using different participants. There were total 3 male nonpaying participants with the age 25, 24 and 22 years respectively. For real time frame analysis, we used Microsoft Visual Studio 2010 and OpenCV 2.4.9. The camera specification is given in the following Table 1.

Table 1: Camera specification

Model	Logitech c170
Photo quality	5 MP
Video quality	HD 720p
Focus type	40 cm and beyond
Auto light correction	Premium

3.2 Data Collection

Data were collection in successive steps which are discussed below-

a) Head and Eye Center Detection

In head and eye pupil tracking experiments, the participants were asked to seat at different distances from the camera under different lighting conditions to set the threshold value. They sat at 30cm, 40cm, 50cm, 60cm, 70cm, 80cm, 90cm and 100cm away from the camera. Different lighting conditions were provided by deploying different number of (1, 2, 3, and 4) 32 Watt energy bulbs. The provided illumination was 50, 100, 150, and 200 Lux respectively (measured by light sensor). The area of our room was $4m^2$. The average video length was 2 minutes.

b) Visual Focus Detection

For gaze detection through applying different techniques, these participants were asked to seat 70 cm away from the camera and to look at different target objects. At first they just moved their eyeball to look at these objects without moving their head. And secondly, they rotated their head to look. These objects are situated 0.5m apart from each other and 1.5m from the target person. One of the three objects is located in the central field of view (CFV) that means just in front of the target person and the other two objects in the near peripheral field of view (NPFV) that means to the left and right side of the target person. The average video recording time was 2 minutes. The provided illumination was 200 Lux.

c) Sustained and Transient Focus of Attention Detection

In sustained and transient attention detection experiments, the participants were asked to look (just move their eyeball) at the left object and the right object for a very short duration so that they may be considered as transients. The duration of transients was also varied (1, 2, 3 seconds) during different experiments. Transients were also detected with the varying head rotation time (1, 2, 3 seconds). The average video recording time was 2 minutes. The provided illumination was 200 Lux. The threshold value for the right distance from the camera was set 70 cm. Figure 2 illustrates some sample frames that were captured during the data collection procedure.

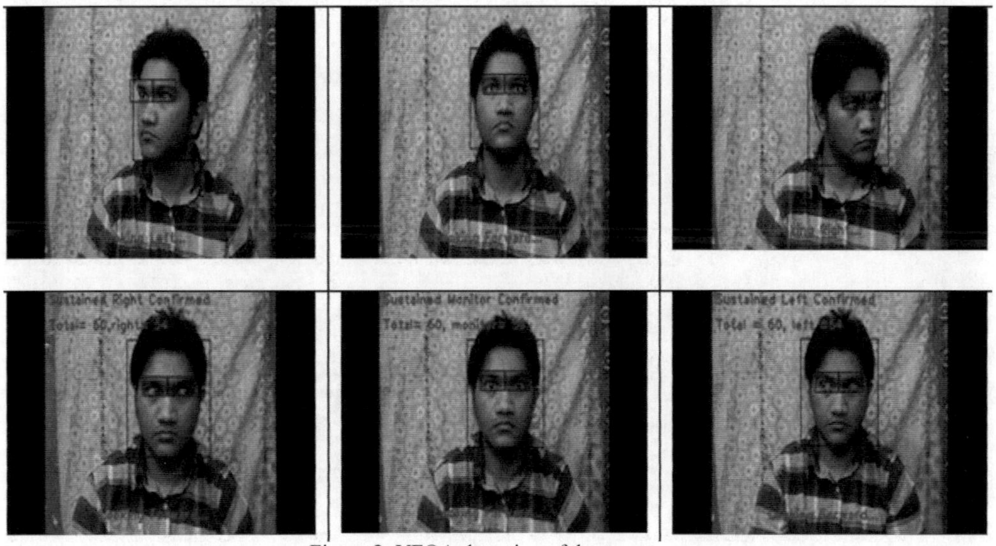

Figure 2. VFOA detection of the target person

3.3 Performance Evaluation Matrices

The efficiency of VFOA tracking under different lightening conditions and different distances from the camera is evaluated based on the following formulae-

Accuracy =

$$\frac{\text{Total number of trials in which head and eye center is detectd}}{\text{Total number of trials at a particular distance from camera}} \times 100$$

%...........(1)

Under the threshold value of right distance from the camera and proper lighting, the efficiency of the transient detection is based on the following formulae-

Transient Detection Accuracy =

$$\frac{\text{Total number of transient attention detected}}{\text{Total number of transient attention occured}} \times 100\ \%..(2)$$

And similarly the efficiency of head rotation detection is expressed as-

Head Rotation Detection Accuracy =

$$\frac{\text{Total number of head movement detected at a particular direction}}{\text{Total number of head movement occured at that particular direction}} \times$$

100%...(3)

4. EXPERIMENTAL RESULTS

Based on Equation 1, we derived the efficiency for head and eye center tracking which are plotted in Figure 3 and 4 respectively. The transient attention detection accuracy is derived based on Equation 2 and shown in Figure 5. Finally, we get the Figure 6 that illustrates the head rotation detection accuracy based on Equation 3. Figure 7 illustrates the transient attention pattern of the target person for typical situations on the time span of about 1 minute.

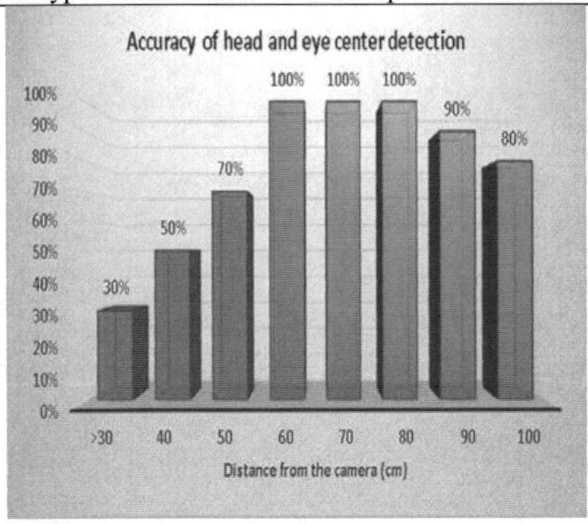

Figure 3. Accuracy of the head and eye center detection with varying distances from the camera

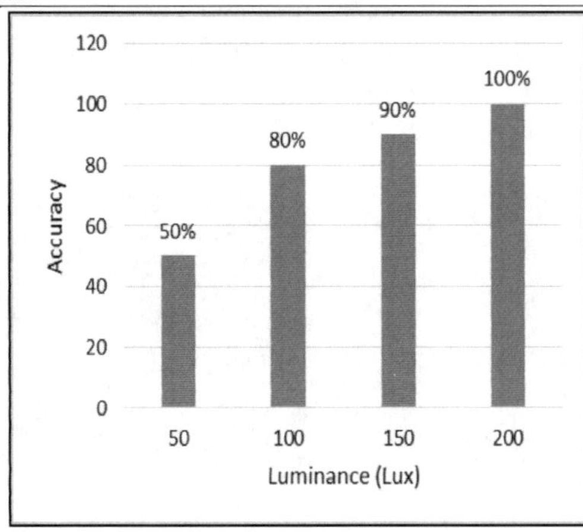

Figure 4. Accuracy of the head and eye center detection with varying lighting conditions

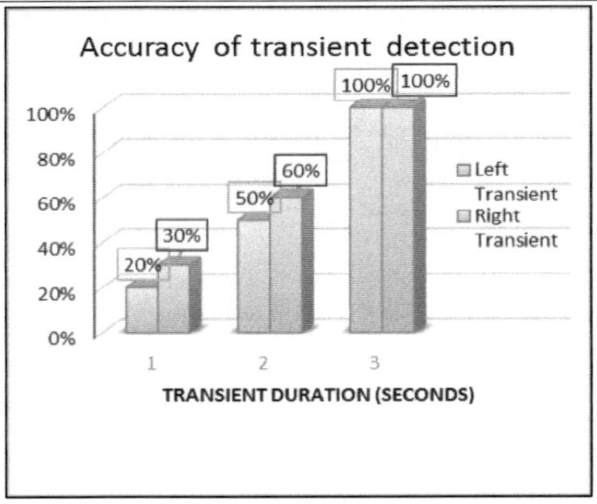

Figure 5. Accuracy of transient detection (eyeball rotation only)

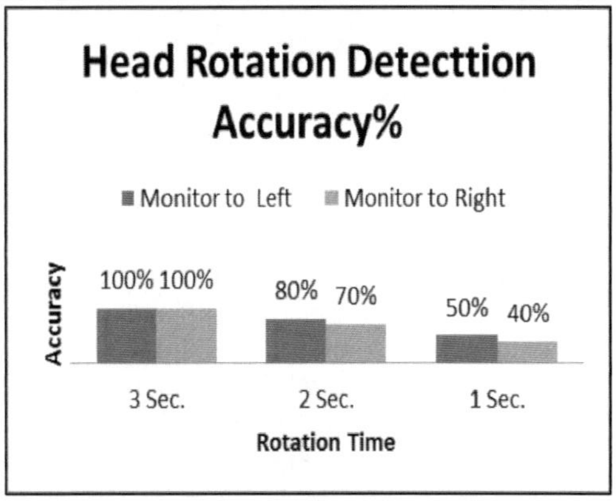

Figure 6. Accuracy of transient detection (head rotation only)

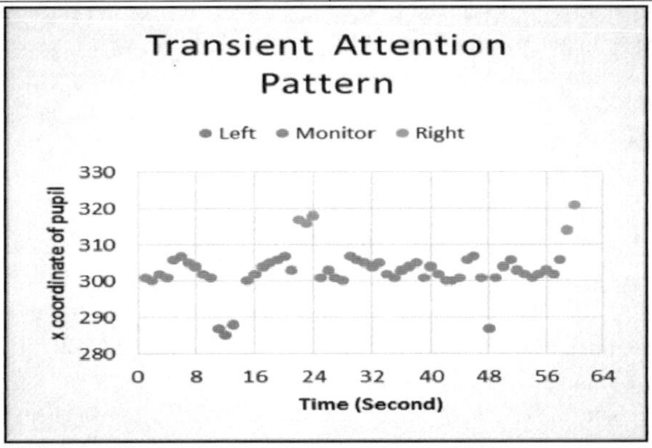

Figure 7. Transient attention pattern of the target person

5. CONCLUSION

In the work, we tried to track the head and the eye center from the captured continuous image with variable distances from the camera and in different lighting conditions. We see that if the distance from the camera is between 60cm to 80cm, we get the best tracking performance. Moreover, for a proper illumination, the number of 32W energy bulb is 4 for a $4m^2$ room that provided illumination of 200 Lux. To optimize the efficiency, this information provides us a clue to set the proper threshold value of illumination and right distance from the camera. In gaze detection combining with head pose we conducted different experiments with variable head rotation time. It is seen that, if the head rotation time is 3 seconds then the gaze tracking performance is the best.

In sustained and transient focus of attention detection experiments, our system can track the sustained attention with 100% accuracy. However, the transient tracking performance depends on the transient duration. We see that if the transient time is 3 seconds or more, the tracking performance is the best.

REFERENCES

[1] Judd T., Ehinger K., Durand F., and Torralba A., "Learning to Predict Where Humans Look," in IEEE 12th international conference on Computer Vision, pp. 2106-2113, IEEE, September 2009.
[2] Alexe B., Deselaers T., and Ferrari, "What is an object," V. 2010. CVPR, pp. 73-80.
[3] Cal H., Lin Y., and Mourant R., "Evaluation of Drivers' Visual Behavior and Road Signs in Virtual Environment," in proceeding of HFES 51st annual meeting, Baltmore, vol. 5, USA, pp. 1645–1649, 2007.
[3] L. Itti and C. Koch, "Computational modelling of visual attention,"Nature Reviews Neuroscience, vol. 2, no. 3, pp. 194–203, Mar 2001.
[4] Zhang H., Smith, M. R., and Witt, G. J, "Identification of Real-Time Diagnostic Measures of Visual Distraction with An Automatic Eye-Tracking System," Human Factors: The Journal of the Human Factors and Ergonomics Society, 48(4), pp. 805-821, 2006.
[5] Md. Golam Rashed, Royta Suzuki, Takua Yenezawa, Antony Lam, Yoshinori Kobayashi, and Yoshinori Kuno," Robustly Tracking People with LIDARs in a Crowded Museum for Behavioral Analysis" Institute of Electronics, Information and Communication Engineers (IEICE) Transactions on Fundamentals of Electronics, Communications and Computer Sciences, Vol E100, No. 11, PP 2458-2469, Nov 2017
[6] Murphy-Chutorian, E., & Trivedi, M. M. (2009). Head Pose Estimation in Computer Vision: A Survey, IEEE Transactions on Pattern Analysis and Machine Intelligence, 31(4), pp. 607- 626.
[7] Kim S., Chung S. T., Jung S., Kim, J., and Cho S. "Multi-Scale Gabor Feature Based Eye Localization," World Academy of Science, Engineering and Technology 21, pp. 483-487, 2007.
[8] Kroon B., Hanjalic A., and Maas S. M., "Eye Localization for Face Matching: Is It Always Useful and Under What Conditions?" in Proceedings of the 2008 international conference on Content-based image and video retrieval, pp. 379-388, ACM, July 2008.
[9] Asteriadi, S, Nikolaidis N, HajduA & Pitas I. (2006, March). An Eye Detection Algorithm Using Pixel to Edge Information. In Int. Symp. on Control, Commun. and Sign. Proc.
[10] B. S and O. J. M, "From camera head pose to 3d global room head poseusing multiple camera views," inIn Proc. Int'l. Workshop Classificationof Events Activities and Relationship, 2007.
[11] An K. H. & Chung M. J. (2008, September). 3d Head Tracking and Pose-Robust 2d Texture Map-Based Face Recognition Using A Simple Ellipsoid Model. In 2008. IROS 2008. IEEE/RSJ International Conference on Intelligent Robots and Systems, pp. 307-312. IEEE.
[12] T. F. Cootes, C. J. Taylor, D. H. Cooper, and J. Graham, "Active shapemodels—their training and application,"Comput. Vis. Image Underst., vol. 61, no. 1, pp. 38–59, Jan. 1995. [Online]. Available:http://dx.doi.org/10.1006/cviu.1995.1004
[13] P. P. Debnath, A. F. M. R. Hasan, and D. Das, "Detection and controllingof drivers' visual focus of attention," in2017 International Conferenceon Electrical, Computer and Communication Engineering (ECCE), Feb2017, pp. 301–307.
[14] David Cornish M. and Dukette D, "The essential 20: twenty components of an excellent Health Care Team," Dorrance Publishing

AUTHOR INDEX